The Brazilian Electricity Market: Small Hydropower Strategic Planning

Thomaz Rodriguez Gonzalez Cortes

ACKNOWLEDGEMENT

As Isaac Newton once wrote, "If I have seen further it is by standing on the shoulders of giants". This thesis is dedicated to the giants that never hesitated to carry me on their shoulders, enabling me to get this far.

Djalma and Carmen, my parents: you are my heroes. Thank you for your infinite support, your invaluable lessons and the innumerous moments of happiness you brought to me. You are, and always will be, my number one source of inspiration, energy and values.

Arthur, my brother, Juliana and Eric, his family: You proved to me, countless times, that we have no limits. You showed that, as Nelson Mandela once said, "It is always impossible until it is done". Arthur, you are an amazing father, a brilliant researcher and the greatest friend.

Muriel, my fiancé: My partner and best friend, thank you for always being unconditionally by my side. Our life will always be adventurous and happy, just like our years in Cambridge.

Professor Robert Pindyck, my advisor: Thank you for your inspiring classes in Industrial Economics for Strategic Decisions, for your brilliant advice and for your invaluable support. It is an honor to work with you.

Gustavo Martinelli Scrignoli, my old friend and electricity market specialist: Thank you for the countless discussions on the Brazilian electricity market. Your support and your vast knowledge on this sector was key to this thesis. I hope you enjoy the reading.

All my friends from the MBA, specially my dearest Brazilian classmates: Thank you for an intense and fun MBA journey together. I hope to see you often in the future, Sloanies!

This page intentionally left blank.

Table of Contents

1. List of Graphs, Figures and Tables

1.1. List of Graphs

1.2.   List of Figures

1.3.   List of Tables

2. List of Abbreviations and Quick Reference Glossary

While reading this thesis, one will soon notice the large use of acronyms and difficult terms through the text. While each concept is explained at the moment it is first used in the document, this section allows the reader to have a quick reference point for each of these words.

ACL	-	Environment of Free Purchasing, allows companies and institutions that consume large amounts of energy to purchase directly from the wholesale market.
ACR	-	Environment of Regulated Purchasing, it is the "regulated market" where distribution companies purchase power on the wholesale to sell on the retail for final users.
ANEEL	-	The National Agency of Electrical Energy, institution responsible for overseeing and regulating electricity in general.
Auto-produtor	-	Self-producer, i.e. an entity that consumes the energy it generates.
A-X	-	A-5, A-3, and so on, are the names of power generation auctions. The name states how many years prior to actual delivery the energy is being negotiated. I.e., in an A-5 energy is sold to be consumed beginning 5 years from the date of the auction.
BRL	-	The Brazilian currency - Brazilian Reais.
CCC	-	A governmental charge based on the consumption of combustibles.
CCEE	-	Chamber of Commerce of Electrical Energy, institution that coordinates the trading of electricity.
CDE	-	A fee used to sponsor development of the electricity sector in certain states.
Eletrobrás	-	A mixed-economy enterprise that invests in electricity generation, transmission and distribution through its several subsidiaries.
EPE	-	Enterprise for Energy Research, a governmental institution responsible for researching and planning the usage of energy in Brazil.
GWh	-	1,000 times larger than one MWh.
Installed Capacity	-	The nominal or theoretical maximum generation flow of a plant, measured in MW.
MME	-	Ministry of Mines and Energy, the part of the federal government responsible for overseeing the usage of resources such as mineral and energy.
MRE	-	The Mechanism for Energy Relocation allows plants to trade their excess/lack of production at the clearing period with a much smaller tariff than the PLD. This tariff is called TEO.
MW	-	Megawatt, a measurement of energy flow (thus only being a measure of consumption after multiplied by time). In analogy to mechanics, MW would be the speed, and MWh the distance traveled.
MWa	-	The average output of a plant in a given period. It is calculated by dividing the energy generation (measured in MWh) by the number of hours of the period. It is also the metric for Physical Warranty.
MWh	-	A measure of energy consumption (see MW).
ONS	-	The National System Operator, responsible for the physical operation of the electricity system in Brazil.
p.a.	-	Abbreviation for Per Annum.

PCH	-	Small Hydropower Plant, with less than 30MW installed capacity. Acronym for Pequena Central Hidrelétrica.
PEN2030	-	It is the long-term plan developed by the Brazilian Government for the energy outlook in the country.
Physical Warranty	-	The average generation expected from a plant, measure in Average Megawatts (MWa). A plant with 1 MWa of Physical Warranty is expected to produce 8760 MWh in a year, as there are 8760h in the year. See section "4.4.1. ACR – "Environment for Regulated Purchase".
PLD	-	Preço de Liquidação das Diferenças, or Price for Liquidation of Differences, the PLD is the "spot price" that players have to pay after at the end of a period if they consumed or sold more energy than bought/produced. It is also the price they receive if they produced more than they sold. It is determined by ANEEL based on the marginal cost of generation in the given subsystem.
Proinfa	-	A fee used by the government for incentivizing clean sources of energy.
SIN	-	The National Integrated System, SIN is the name for the interconnected electrical system that serves the vast majority of the Brazilian Electricity Market.
TEO	-	The TEO is the Tariff utilized by participants of the MRE to trade at the clearing of each period. It substitutes the PLD for differences in actual generation from the Physical Warranty. It has always been lower than BRL 15, while the PLD varies from BRL ~30 to ~388.
TUSD	-	The tariff that pays the costs of distribution.
TUST	-	The tariff that pays the costs of transmission.
TWh	-	1,000,000 times larger than one MWh.
Utilization Rate of the Plant	-	Also called Power Factor, it is the Physical Warranty divided by the Installed Capacity of a given plant.
WACC	-	Weighted Average Cost of Capital.

3. Introduction and Methodology

This thesis is aimed at providing Brazilian Electrical Energy investors, analysts, traders, students and other professionals with relevant market insights on the field of Small Hydropower generation and the overall Brazilian generation landscape. This thesis is a self-contained document, meaning it provides comprehensive content that enables even a first interaction with the topic. However, other publications could provide more granularity on the Brazilian Electricity Market. A great point of start would be with the websites of the Chamber of Commerce of Electrical Energy[i], the National Agency of Electrical Energy[ii], the National System Operator[iii] and the Enterprise for Energy Research[iv]. Other articles and books may provide better perspectives on the history and the in-depth works of electricity market regulations, such as the comprehensive book "Regulation of the Power Sector" edited by Professor José Ignacio Pérez-Arriaga (Pérez-Arriaga, 2013), or articles like "A Perspective of the Brazilian Electricity Sector Restructuring: From Privatization to the New Model Framework" (Melo, Neves, da Costa, & Correia). The latter provides valuable information on the history of the Brazilian market regulation and its transition to the current paradigm, which has been functioning since 2004.

For a clear organization of the information presented, the thesis is divided in three parts. Aside of these three parts, the reader may find references at the end of this document. The overall structure of the document can be more easily observed in Figure 1.

The first part of the thesis provides an overview of the Brazilian Electricity Generation Market. Such overview includes insights on the current drought and its effects on the electricity supply and prices, as well as observations on the supply expansion plans and routes that the country has taken to follow economic and consumption growth. In the last section of this first part, the reader may find a few observations on investment behavior. This part also explains the

---

[i] http://www.ccee.org.br
[ii] http://www.aneel.gov.br
[iii] http://www.ons.org.br
[iv] http://www.epe.gov.br

basic functioning of the market, allowing the reader to have this thesis as a self-contained document – i.e. being able to fully understand its propositions from itself.

There are two main conclusions of this initial part of the thesis. First, that the actual consumption growth has been and is expected to remain much slower than the governmental plans of electricity supply. The major long-term government plan was designed in 2007, and as Graph 19 will show, the GDP actual growth was far below their worst-case scenario consideration. Indeed, the Brazilian Macroeconomic scenario of 2007 was great, with real growth of 6.1% on that year. Furthermore, Graph 18 shows how this optimistic growth reflected in annual capacity additions. This, together with the expected recovery from the drought of the 70% of national supply that is hydropower-based, points towards a potential oversupply in the mid- to long-term. Second, that the many options of electricity sales provide options for different risk appetites, and investors have utilized them in ways that vary from selling 30 years of future generation at a price fixed prior to construction, to selling all its generation on the month-to-month clearing/spot market with highly volatile prices.

The second part dives with some more depth on hydropower investments – the differentiation of large scale projects to Small Hydropower, capital requirements, generation cycle, value creation levers and risks and options of energy sale. The most interesting insight of this part is that Small Hydropower Plants, defined as hydropower plants smaller than 30MW in installed capacity, have significant advantages when compared to the Large ones (up to 14,000 MW), despite the lack of scale.

Such advantages are due to many factors. First, the learning curve of building several plants, as opposed to a single large one, allows investors to optimize their investment's performance. This happens in three dimensions: Capital Expenditure reduction, actual generation projection, and revenues maximization. Interviews with investors revealed that while the first plant may cost up to BRL 12 Million per MW of installed capacity, this can be reduced after 5-7 plants built to below BRL 4 Million per MW. On top of the investment reduction, there is also a reduction in risk due to the improvement in the precision of the prediction of the actual generation as a percentage of the nominal installed capacity. In other words, for instance, investors with little project experience may build a plant with 10 MW installed capacity expecting to have an 8MW average output, only to discover, after construction, that the actual generation is merely at a 5MW average. Lastly, there is also a learning curve on trading – investors have reported to improve significantly their performance in terms of average sales price per MWh (Small Hydropower Investors, 2015). While new investors may opt for longer-term, lower-risk auction sales, more seasoned investors are able to trade on the free and the spot markets as well.

The second advantage is that Small Hydropower Plants are free to sell the entirety of their generation in any way they want, while the large plants have to sell a minimum of 70% on the lower-price regulated market. Also related to the auction process, the third advantage is that there is no competition to build a single Small Hydropower Plant, while the large plants are built by the company or consortium of companies that commits to sell at the lowest price at the regulated market. This lowest-price bidding process not only reduces drastically the average price per Megawatt-hour of the plant, but also wastes the non-trivial investments in project preparation of the companies that do not win the auctions. Lastly, the fourth advantage is that Small Hydropower Plants produce environmental impacts orders of magnitude smaller than their larger counterparts, due to the size or even the lack of reservoirs. As a study performed by McKinsey and Company showed, Small Hydropower can even produce positive environmental impacts, as they reduce the amount of carbon emissions at each dollar invested (McKinsey and Company, 2014).

Finally, this sets ground for the third and last part of the thesis, which explores sales options and compare them in terms of risk and portfolio management. As a simple closure of this last part, there are demonstrations of how different portfolios have very different outcomes in terms of expected present value of generated revenues and its standard deviation. Value creation levers that will not be explored are such as financial leveraging, tax reduction/shields, operational improvements, increases in efficiency of construction techniques or even reduction in construction time, although each one of them is highly important and could significantly impact the return on investment of hydropower plants.

The most important insights of this last part are related to the three variations applied to the model created for this thesis. The first insight is that the standard deviation of the monthly generation, which is given by both the construction of the plant and the water flow, could cause large deviations in the present value of revenues of a plant. Furthermore, such variations are much larger to shorter-term portfolios and those not part of the MRE, a regulated pool that allows plants to trade excess/lack in capacity at very low prices. The second insight is that the future length of the current drought may impact significantly the expected present value of revenues of a given plant. Portfolios more exposed to spot and free market prices are even more affected by the drought, while long-term portfolios with higher share sold at the regulated market are much more stable and independent of such variable. Lastly, the third insight is that players that chose to sell their whole generation of 30 years at prices fixed prior to construction of their plants have very low standard deviations of revenues. On the other hand, the expected value in some scenarios might be low enough that, even with higher deviations, shorter-term scenarios might be interesting as well. Finally, this part of the thesis demonstrates the extent of the impact, and thus the relevance of portfolio-thinking on the electricity market. However, it does not provide tools to enable investors to effectively optimize their portfolio based on their risk appetites. Therefore, as mentioned in section "7. Conclusion and Further Studies" ahead, the actual optimization of the portfolio of electricity sales is a very interesting field for further research.

In each one of the three parts there was a careful focus on maximizing the usage of support data and reliable sources. As it may be noted in the footnotes and references, there is a quite large and fairly decentralized amount of data on the Brazilian Electricity market. Some sources, however, are individually very organized, complete and detailed, which facilitates the mathematical analysis and the derivation of relevant insights. A good example of this can be seen in CCEE's InfoMercado (CCEE, 2015) – a report that details actual generation in each one of the active power plants in the market. Another great example is the National System Operator's report on water flows, which covers 194 key points of hydropower generation within the period of 1931 to 2013. On the other hand, other sources may be very inadequate. The most negative of them is the absolute lack of data of prices and terms of energy contracts in the free market.

CCEE was authorized in 2012 to release indexes of prices for the free market (Canal Energia, 2015) through the "Portaria 455", but before they could actually do so electricity agents managed to utilize judicial means to prohibit such publication.

The Scope of the Thesis

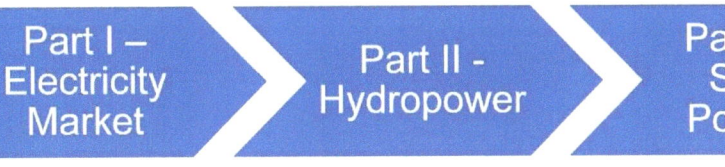

Part I – Electricity Market
- Sources and Usage
- Regulatory Structure
- Energy Trading
- Future Expectations
- Investors' Behavior

Part II - Hydropower
- Large vs. Small Hydropower
- Investing Process
- Capacity and Physical Warranty
- MRE Pool Effects
- Role of Hydropower in the Brazil

Part III – Sales Portfolio
- Decisions to Build the Portfolio
- Revenue Sources
- Modelling Structure
- Simulations of Present Value of Revenues
- Modelling Insights

Figure 1 - The Scope of the Thesis

4. Part I – The Brazilian Electricity Generation Market

4.1. Current Sources of Electricity

The Brazilian Electricity market is the 9th largest in the world, as shown in Graph 1, having consumed 557 Terawatt-hours of energy in 2013 (BP Global, 2015). With close to 70% of its installed capacity coming from Hydroelectric Power Plants (EPE, 2014), Brazil is the 3rd largest hydropower consumer in the world (BP Global, 2015), as shown in Graphs 3 and 4. The country has been growing steadily its installed capacity at a compound rate of 5.1% per year in the period of 1974-2013, with the recent years showing a growth of 2 to 7% year-over-year (EPE, 2014).

As an interesting note, Brazil has a quite low average consumption per capita, as shown in Graph 2, even while having 85% of population in Urban Areas. Korea, for instance, has 82%, according to the World Bank (The World Bank, 2015), and still has almost four times the per capita consumption. However, 40% of "economically active" population, as defined by the Brazilian Institute of Geography and Statistics, IBGE, lives with less than the minimum wage of BRL 788 (USD ~260) monthly salary (or USD 3,500 per year). This also excludes that fact that the "economically active" population accounts for only about 60% of the total population (IBGE, 2010). On top of that, there are much lower penetration rates of household appliances, there is no need for heating, only 13% penetration rate of Air Conditioners (ASBRAV, 2014), and cooking is almost always done with gas and not electricity.

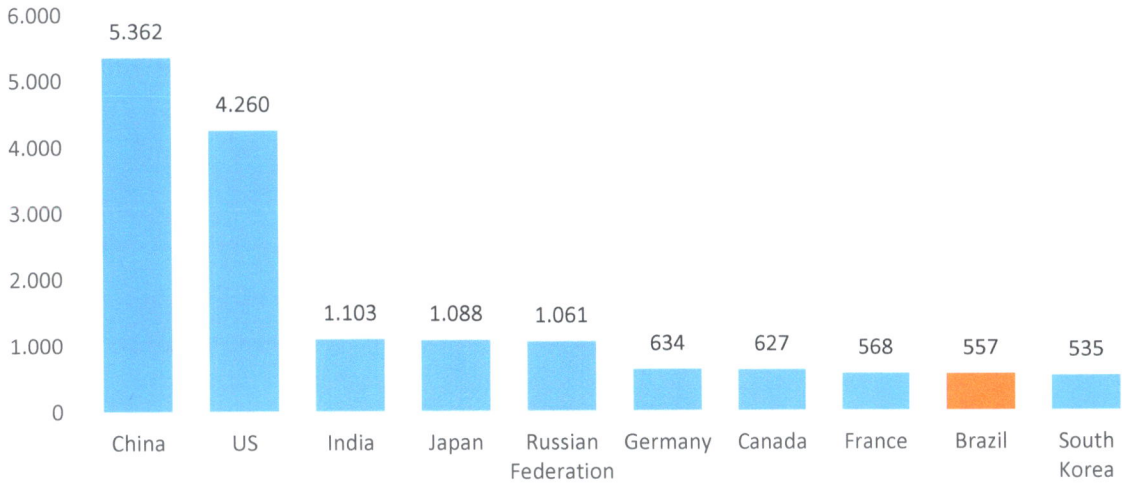

Graph 1 - Top 10 Countries by Electricity Consumption in Terawatt-Hours

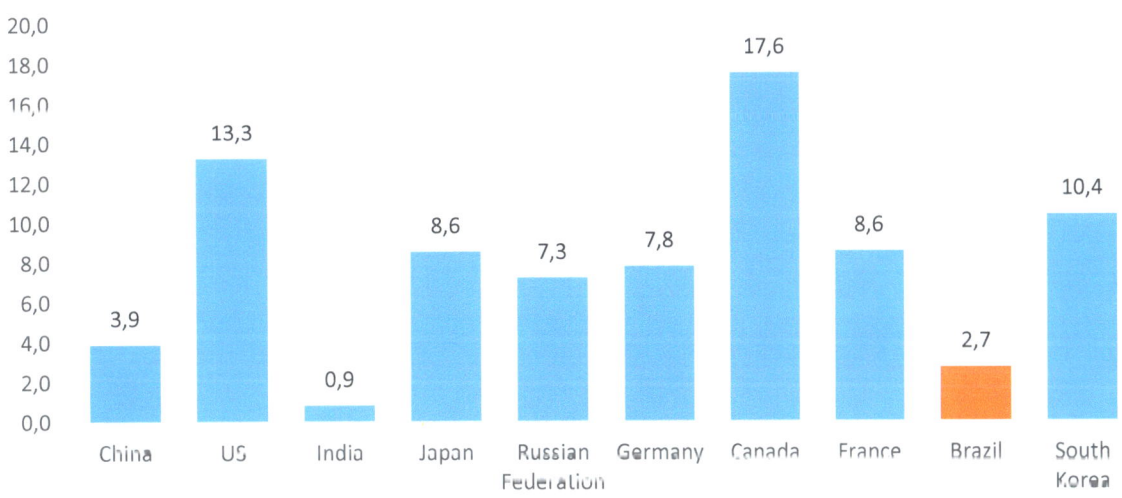

Graph 2 - Top 10 Countries by Electricity Consumption in Megawatt-hours per capita

To diminish the dependency on hydropower demonstrated by the breakdown in Graph 5, and thus to reduce its vulnerability to weather conditions and such uncontrollable factors, the Brazilian electricity market has been investing strongly in expanding its capacity through other sources. While the Compound Annual Growth Rate of the period 2003-2013 was 2.4% for hydropower installed capacity, this is mainly due to strong investments in self-generation[v]. Within hydro, this type of generation grew 14.8% versus 2.0% of market-focused producers. As explained in more detail in the section "4.4.5. Self-production – the role of the 'Auto-Produtor'", self-generation plants are owned by companies that also consume the plant's production, thus not being sold through the free or the captive market directly.

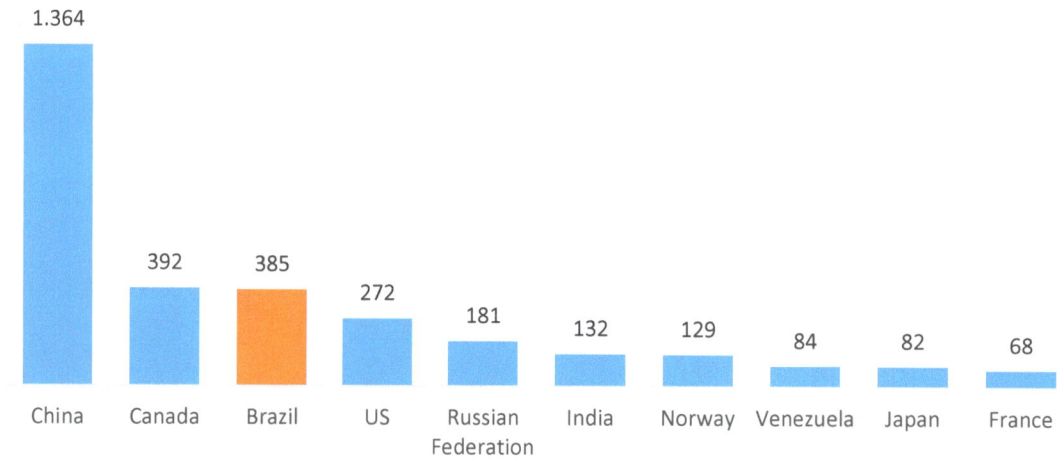

TOP 10 COUNTRIES BY HYDROPOWER
Annual consumption of Hydropower in Terawatt-hours, 2013

Source: BP Statistical Review of World Energy 2014

Graph 3 - Top 10 Countries by Hydropower in Terawatt-hours

---

[v] Self-generation is defined by power plants owned by large electricity consumers, as detailed in the section "3.4.5. Self-production – the role of the 'Auto-Produtor'"

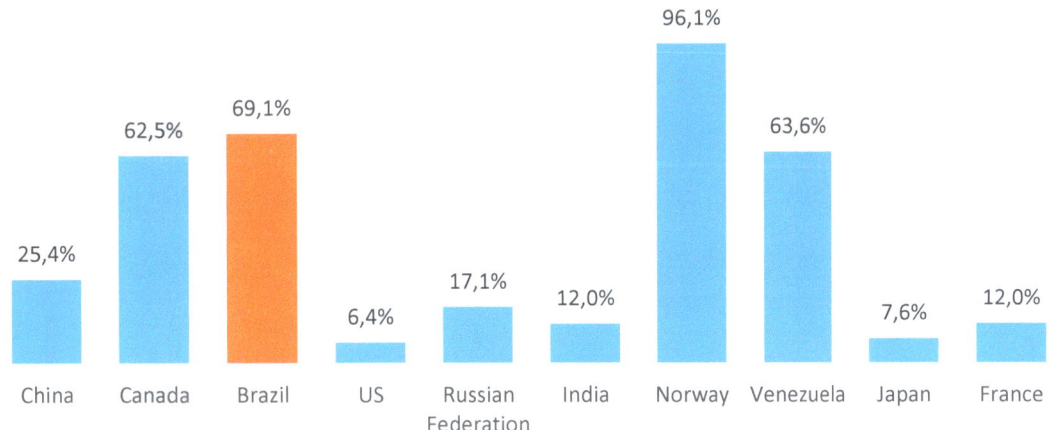

TOP 10 COUNTRIES BY HYDROPOWER
Annual consumption of hydropower as share of total consumption,
2013

China 25,4% Canada 62,5% Brazil 69,1% US 6,4% Russian Federation 17,1% India 12,0% Norway 96,1% Venezuela 63,6% Japan 7,6% France 12,0%

Source: BP Statistical Review of World Energy 2014

Graph 4 - Top 10 Countries by Hydropower in Percentage of Total Electricity Generation

A similar trend of faster increase in self-production affected the Thermal sources, such as coal or oil. While the Thermal generation grew at 8.5% CAGR in 2003-2013, its self-production grew 12.1%, versus a 6.6% increase in capacity with a market focus. By December 31st 2014, 35% of the Thermal capacity was based on Gas, 33% on Biomass, 23% on Oil, 9% on Coal and 1% on other sources (ANEEL, 2015).

From 2005 on, Wind-powered generation began to have significance in the Brazilian market as well, growing from 22 MW to 2.2 GW in 2013 (EPE, 2014). As briefly cited in section 4.5., Wind generation in Brazil could be interestingly complementary to hydropower, which is could be useful for companies with more exposed portfolios. Almost the totality of the Wind generation Is focused on market and not on self-production, mainly due to subsidies that the government provide to this power source[vi]. Both Thermal and Wind power are the only relevant efforts currently being undertaken by the market to reduce its dependency on hydropower.

[vi] Most subsidies in the electricity market are given via reduction in fees and taxes, such as TUST (Transmission fees) and Proinfa exemptions

Lastly, Solar Power is still in its infancy in Brazil - yet at less than 0.001% of share of capacity, with only 15 MW installed out of the 133,913 MW total by December, 2014 (ANEEL, 2015). Nuclear power, on the other hand, is only available through Angra, a single cluster of three plants located in the southeast of the country. Angra I and Angra II are operational and have about 2 GW of installed capacity. Angra III was a project with 1.4 GW of capacity from 1984 that was halted, but it has recently been resumed with plans to enter the market by 2018. The government had plans to build four more nuclear power plants, but due to the Fukushima accident, its nuclear plans were cancelled and are not expected to be resumed at any point in the future, as declared by EPE's president, Maurício Tolmasquim (Brazilian Federal Government, 2011).

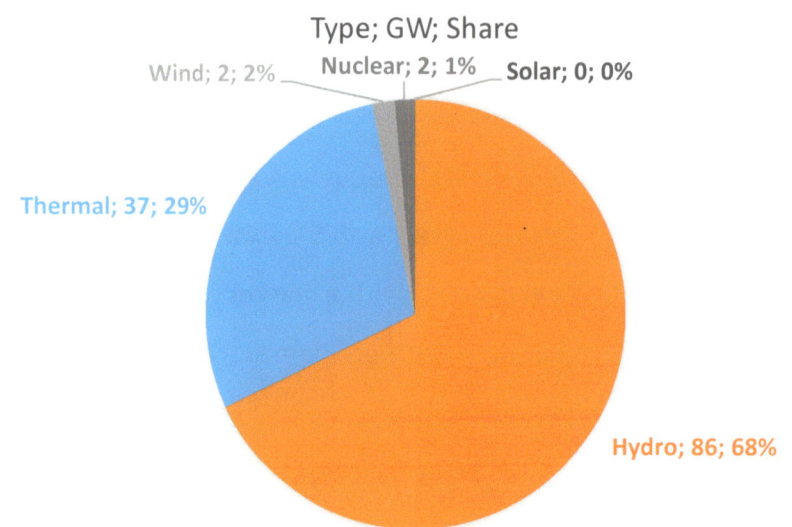

INSTALLED GENERATION CAPACITY - BY SOURCE
Type; GW; Share

Wind; 2; 2%   Nuclear; 2; 1%   Solar; 0; 0%

Thermal; 37; 29%

Hydro; 86; 68%

Source: Balanço Energético Nacional 2014 - EPE

Graph 5 - Installed Generation Capacity in Brazil by Source

4.2. Electricity Usage and System

4.2.1. The SIN - Sistema Interligado Nacional – and its Regional Subsystems

The Brazilian Electricity System, named "SIN", is a unified system that connects 98.8% of the electrical energy consumed in the country. Apart from small isolated systems that remain disconnected due to the immense territory of the country, the SIN allows the transmission of energy to and from almost any region of the country. The SIN system is divided into four subsystems, based on geographical proximity and concentration of consumption as Figures 2 and 3 illustrate. As it can be seen on Graph 6, the south and southeast subsystems account for almost 80% of the consumption and hold the majority of the hydropower generation. As the current drought affected most strongly the southeast region, it becomes clear why its impact is so relevant for electricity generation. Such impact will be further explored in section "4.5. Expansion Plan, Economy Growth and the Drought".

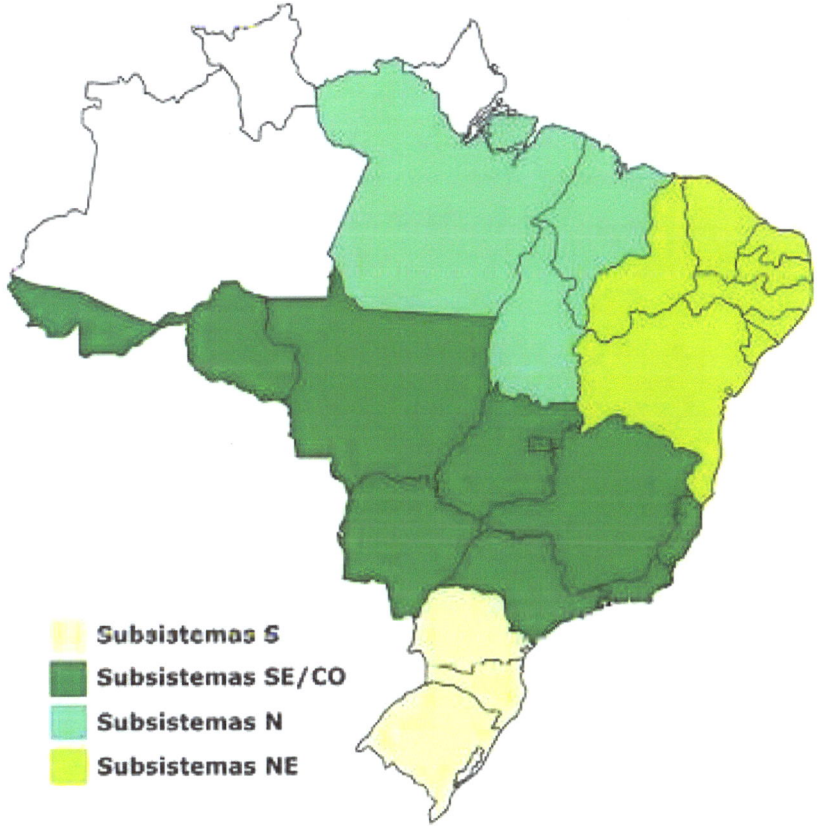

Figure 2 - The SIN Sub-systems; Source: ANEEL

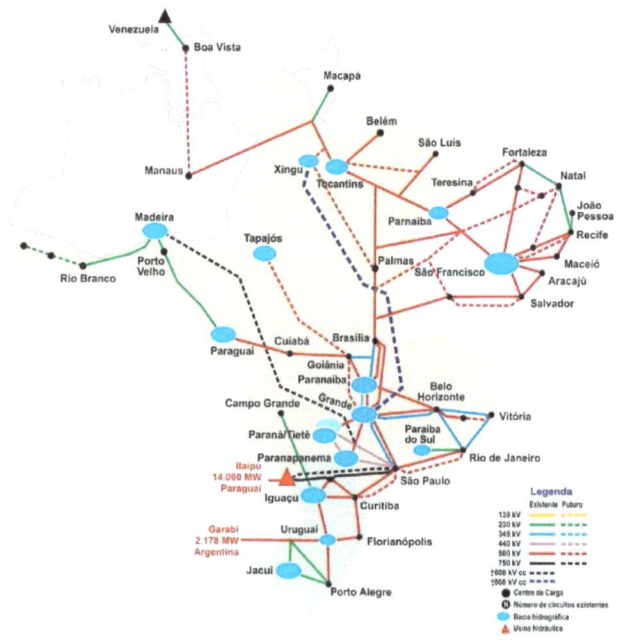

Figure 3 - The SIN's Transmission lines and connectivity; Source: ONS

ELECTRICITY CONSUMPTION - BY SUBSYSTEM
Subsystem; TWh; Share

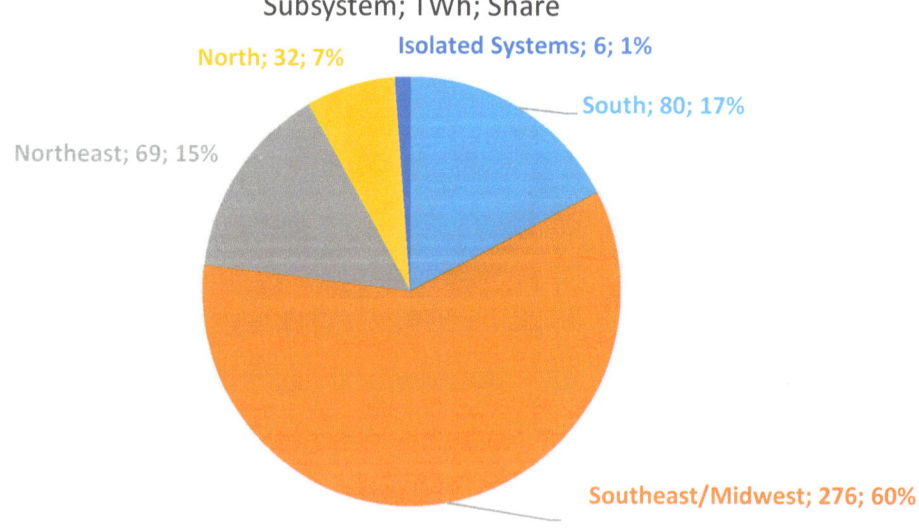

Source: Anuário Estatístico de Energia Elétrica 2014

Graph 6 - Electricity Consumption by Subsystem

4.2.2. Final Uses of Electricity

As detailed in Graph 7, electricity in Brazil is 40% consumed by industry, with Residential and Commercial usages representing 45% of the total (EPE, 2014). As a quick comparison, electricity usage by the industrial sector in the United States represents only 25% of total consumption, with residential and commercial each representing half of another 75% of consumption, according to the 2013 report of U.S. Energy Information Administration (EIA-USA, 2015). As detailed ahead, the large industrial usage has implications for the electricity suppliers, as industrial consumers comprise the vast majority of the purchases of free market energy.

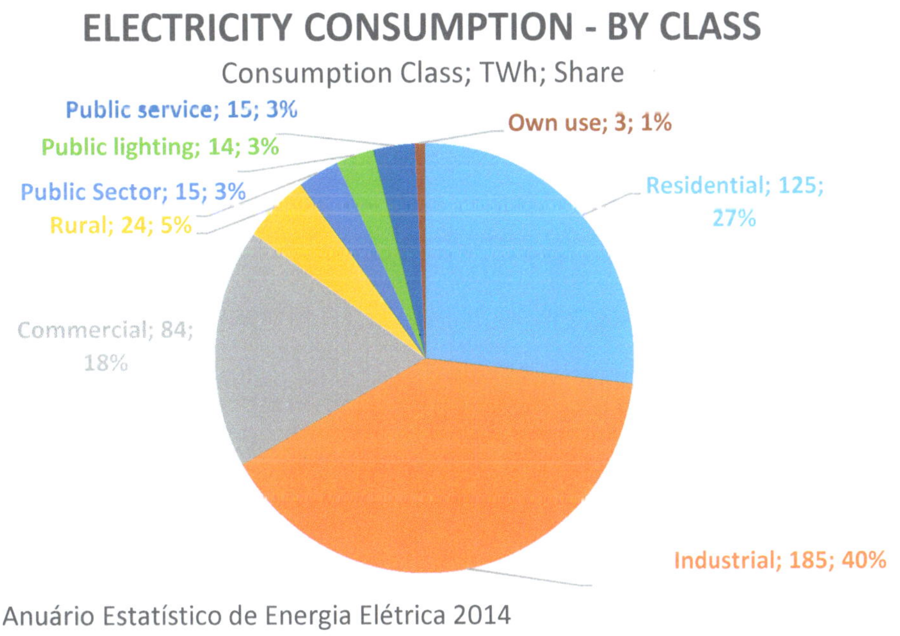

ELECTRICITY CONSUMPTION - BY CLASS
Consumption Class; TWh; Share

- Public service; 15; 3%
- Public lighting; 14; 3%
- Public Sector; 15; 3%
- Rural; 24; 5%
- Own use; 3; 1%
- Residential; 125; 27%
- Commercial; 84; 18%
- Industrial; 185; 40%

Source: Anuário Estatístico de Energia Elétrica 2014

Graph 7 - Electricity Consumption by Class

4.3. The Regulatory Structure: MME, Eletrobrás, CCEE, ONS, ANEEL and EPE

Although it is far from the goal of this thesis to be a detailed guide to the complex Brazilian electricity regulatory framework, it would be unreadable as a self-contained document should it not explain the roles of the main regulatory institutions and the key concept behind its regulation. The six main institutions are briefly described here, and should be enough to understand the sources of this thesis as well as to allow the reader to have some initial direction in the search for more information.

The institutions listed below have very well-defined roles that affect the electricity market, and as they are all part of the Brazilian government, they act in a cooperative way. The MME is the ministry that responds directly to the president, and it acts mostly by defining the roles of all the other institutions mentioned here. ANEEL is the agency that effectively regulates the electricity generation, transmission and distribution. EPE is a "Management Consulting arm", responsible for information analyses and long-term planning. The CCEE oversees trading, keeping information on all sales, including bilateral agreements, as well as clearing the market at the end of each period. The ONS is the physical operator of the system, controlling dispatch of plants and transmission of electricity. Lastly, Eletrobrás is a mixed-economy company that invests in all segments of the electricity market.

4.3.1. MME – Ministério de Minas e Energia[vii]

Founded in its current format in 1992, by the law 8.422 issued by the Brazilian government, the Ministry of Mines and Energy is the leading institution on the areas of geology, mineral resources, energy (including all sources of electricity), steel, oil, combustibles and renewables. The current structure of the Ministry includes the Secretaries of "Planning and Energy Development", "Electrical Energy", "Oil", "Natural Gas and Renewable Combustibles", "Geology, Mining and Mineral Transformation" (MME, 2015). The autarchy ANEEL, the National Agency for Electrical Energy, is directly linked to the MME, and in 2004, by the law 10.847, the government authorized the MME to create the "Empresa de Pesquisa Energética", or EPE – two of the key institutions described here. Lastly, the mixed economy (i.e. partly state-owned) companies Eletrobrás and Petrobras are both directly linked to the MME as well.

4.3.2. ANEEL

ANEEL stands for "Agência Nacional de Energia Elétrica", or "National Agency of Electrical Energy". Its mission is to "provide favorable conditions that enable the electrical energy market

_____

[vii] Text largely based on MME own website.

to develop with balance among the agents and in benefit of the society". ANEEL responds to the MME, and is responsible from issuing the "rules of the market" (ANEEL, 2015), from the documents that detail rules of each specific auction to the definition of the spot prices and evaluation of the market performance. In other words, ANEEL is the "executive" branch of the Brazilian Electricity Market, carrying out governmental plans and ensuring that the market is balanced and developing at the right pace. Furthermore, ANEEL detains a large part of the regulatory power, being responsible for the content of the laws through which the government regulates the energy market. Although the government as an institution already has the power to create, exclude and modify the laws that regulate the market, it does so through ANEEL when it comes to generation, transmission and distribution of electricity.

4.3.3. EPE

EPE stands for "Empresa de Pesquisa Energética", or "Enterprise for Energy Research". In summary, its function is to be the "Management Consulting arm" of the government when it comes to energy in general, responding directly to the MME. Its function is to "provide services in the area of studies and researches with objective of subsidizing the planning of the energy sector, such as electrical energy, oil, natural gas and its derivatives, mineral coal, renewable energy sources, among others" (EPE, 2015). The EPE main task is to develop and maintain the national energy long-term plan, while evaluating its current and past performance.

4.3.4. CCEE

The "Câmara de Comercialização de Energia Elétrica", or "Chamber of Commerce of Electrical Energy", has the goal of "enabling the activities of buying and selling energy in Brazil". Its main function is to oversee the trading of energy. While it does provide valuable insights and information to ANEEL, it does not have direct regulatory power. It also "promotes discussions regarding the evolution of the market, always oriented by the pillars of equality, transparency and trustworthiness" (CCEE, 2015). The CCEE coordinates 2,915 agents, including generators, distributors, consumers, self-producers and traders. It has already coordinated 58 auctions, while registering 20,524 contracts of energy (CCEE, 2015). Every contract of purchase of electricity has to be done through the CCEE, whether they are done through auctions (ACR, discussed in section

"4.4.1. ACR – "Environment for Regulated Purchase") or bilaterally (ACL, detailed in section "4.4.2. ACL – "Environment for Free Purchase").

4.3.5. ONS

The ONS "Operador Nacional do Sistema Elétrico" is the state-owned, centralized operator of the national electricity system – the SIN. While ONS does not play a direct role in the economic side of the electricity market, it is responsible for acting on the physical system, from controlling power transmission to measuring outputs and dispatching power plants. Most of the large power plants and many of the smaller ones are centrally dispatched, i.e. they provide electricity to the market whenever the ONS decides it is optimal for the system. Due to the large hydropower share in the Brazilian supply, the ONS also keeps close track of the water flows and the reservoir levels in the country. Furthermore, many of the hydropower plants are cascaded on single water flows, meaning that an optimal, centralized dispatch is necessary to ensure that the upstream plants do not utilize water flow as a market power lever.

4.3.6. Eletrobrás

As described above, Eletrobrás is a mixed economy enterprise that focus on investing in the electricity market in Brazil. In other words, Eletrobrás is effectively a company that invests in energy generation, transmission and distribution. Furthermore, there is a strong presence of Eletrobrás in the ownership of several power generation facilities, even more so when it comes to very large plants with structural importance for the country.

Even within the current market regulatory framework (since 2004) that allowed private companies to bid to build even the largest plants, the main hydropower facilities (with more than 200MW capacity each) have around 40-50% ownership of at least one Eletrobrás subsidiary. This phenomenon can be seen, for instance, in the Hydropower plants of Dardanelos (261MW; 49% Eletrobrás), Mauá (362MW; 49% Eletrobrás), Teles Pires (1.820MW; 49% Eletrobrás), Santo

Antonio (3.150MW; 39% Eletrobrás), Jirau (3.300MW; 40% Eletrobrás) and Belo Monte (11.233MW; 50% Eletrobrás) (ANEEL, 2004-2015)[viii].

Although a very interesting matter, possibly enough for a whole other thesis, it is not the goal of this document to debate whether this is an efficient manner of having the government to subsidize the return of other companies in such consortiums. However, the BNDES (National Bank of Development) funding of usually 70% of the capital utilized in such plants, together with a few very low expected returns by Eletrobrás, could be indications of such practice. Such low returns may be observed in plants like Jirau, Santo Antonio and Belo Monte, to which a report made by JP Morgan attributed expected returns of 3.2%, 5.9% and 7.2% p.a. respectively (Estado de São Paulo, 2014).

Lastly, Eletrobrás acts mostly through its controlled subsidiaries, namely the companies Furnas Centrais Elétricas S.A., Companhia Hidroelétrica do São Francisco (Chesf), Companhia de Geração Térmica de Energia Elétrica (CGTEE), Centrais Elétricas do Norte do Brasil S.A. (Eletronorte), Eletrosul Centrais Elétricas S.A. (Eletrosul) and Eletrobrás Termonuclear S.A. (Eletronuclear).

4.4. The Energy Trading in Brazil

In the current Brazilian electrical power regulatory framework, there are mainly three ways for a generator to sell energy. They are named the ACR, the ACL and through self-production, as detailed in this section.

4.4.1. ACR – "Environment for Regulated Purchase"

The ACR, or "Ambiente de Contratação Regulada", allows generators to sell their productions through auctions organized by CCEE under the control of ANEEL. Prior to the auctions, generators register with their plants and the amount they wish to sell, and energy is purchased by buyers – distribution companies – who decide based on the cheapest price per Megawatt-hour available and the quantity they need to supply their end-consumers. At the act

---

[viii] Sourced from ANEEL and the reports released at the conclusion of the auction where each of the cited plants participated.

of purchase at the auction the price is defined for the whole period of supply, only to be readjusted by official inflation metrics. All the energy consumed by the residential sector, and about 73% of the whole electricity consumed in the country in 2013 was traded in the ACR.

Most of such auctions are named in the pattern "A-X", where X is the nominal number of years prior to the actual delivery of energy. An auction that happens in any day of 2015 to start delivering in January 1st, 2016, would be called an "A-1", and so on. Auctions that are A-3 and A-5 are often utilized to allow investors who will yet begin building to have assured revenues after the plant begins operating, which enables them to get better financing than otherwise. In the case of large hydropower (over 30MW of capacity, sometimes even larger than 10,000MW), different companies and/or consortiums compete to build the same plant, by offering the price at which they commit to sell energy at the ACR should they win the bid. In this situation, the lowest price offered wins the right to build the plant.

Analyzing the 937 plants that sold at least a part of their generation in the ANEEL auctions, it is found that approximately 90% of them, representing about 83% of the total auctioned energy, sold to begin supplying after 3 years or more, thus showing the importance of the sale prior to construction in the Brazilian market (CCEE, 2004-2015).

Within Small Hydropower generators (30MW or less in capacity, also known as "PCHs"), the focus of this thesis, there were 48 different plants that participated in auctions, since the beginning of this regulatory framework in 2004. Each of these plants participated only once – 45 of them selling energy prior to construction for a period of 30 consecutive years from the date of beginning of operations. Of these 45, 14 sold in A-3 auctions, while 30 sold in A-5 auctions, and one did not have this data available (CCEE, 2004-2015). This breakdown can be seen on Graph 8.

PARTICIPATION OF PCH IN AUCTIONS - A-X CHOICE

A-X; Number of Plants; Share

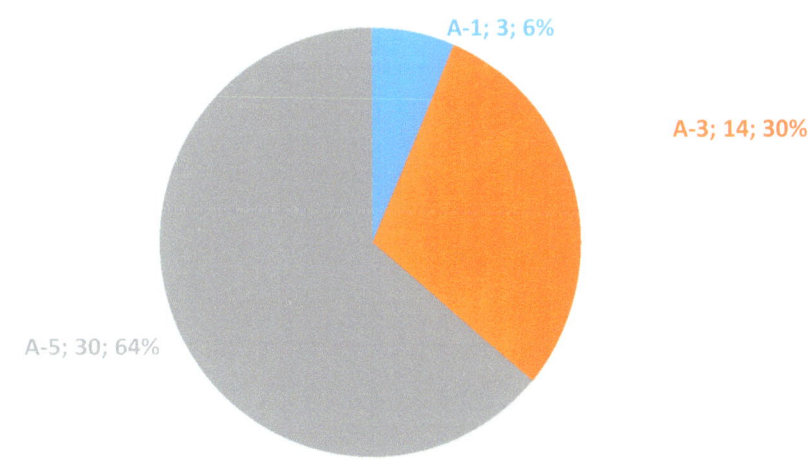

Source: CCEE - Resultado Consolidado de Leilões - February/2015

Graph 8 - Participation of PCHs in Auctions by A-X Choice

Both A-3 and A-5 allow the PCH investors to commit to sell energy for a period of up to 30 years at a price specified at the date of the auction and readjusted by inflation, thus allowing future revenue to be leveraged in order to obtain better finance 3-5 years prior to construction. As it can be seen on the Graph 9, the majority of plants that choose this type of long-term sale include most of their generation in the auction. Furthermore, PCHs are not eligible to participate in A-3 or A-5 auctions once they have been built.

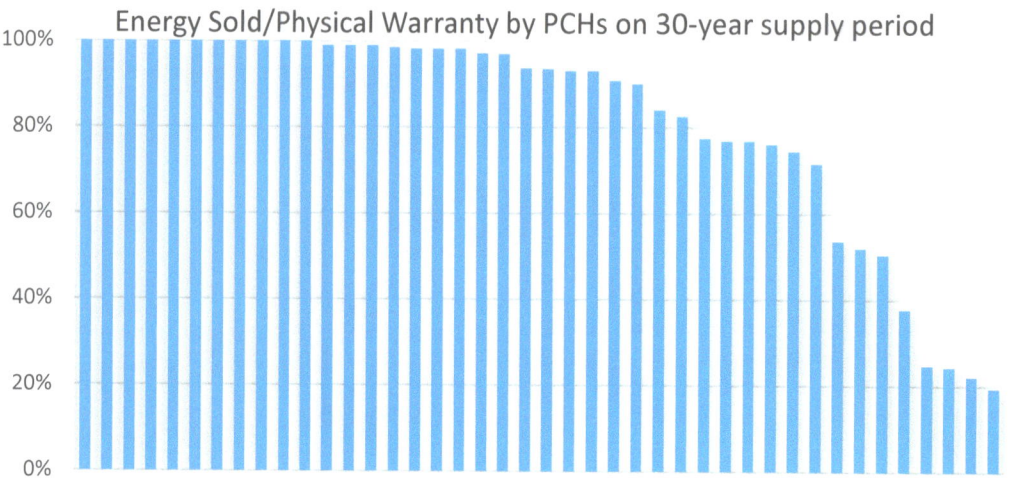

PCH IN AUCTIONS - SHARE OF PHYSICAL WARRANTY SOLD

Energy Sold/Physical Warranty by PCHs on 30-year supply period

Source: CCEE - Resultado Consolidado de Leilões - February/2015

Graph 9 - PCH in Auctions by Share of Physical Warranty Sold

The other 3 plants sold energy supply at A-1 auctions, for a period of 5 years. PCHs can participate in A-1 auctions at any given point in their lifetime, and may sell energy, in theory, for a maximum of 8 years.

Lastly, in any auction each plant may sell any share of its production, up to 100% of its Physical Warranty. The Physical Warranty is the effective production capacity, measured in Average Megawatts (MWm in Portuguese, or MWa in English). It is defined by ANEEL, based on physical measurement, project evaluation and history performance – in other words, ANEEL defines the expected production based on all the data it can gather from the plant. A brief explanation on the relationship between Installed Capacity and actual production can be found in the Figure 4 below. More on the utilization rate of Small Hydropower plants can be found in the third part of this thesis.

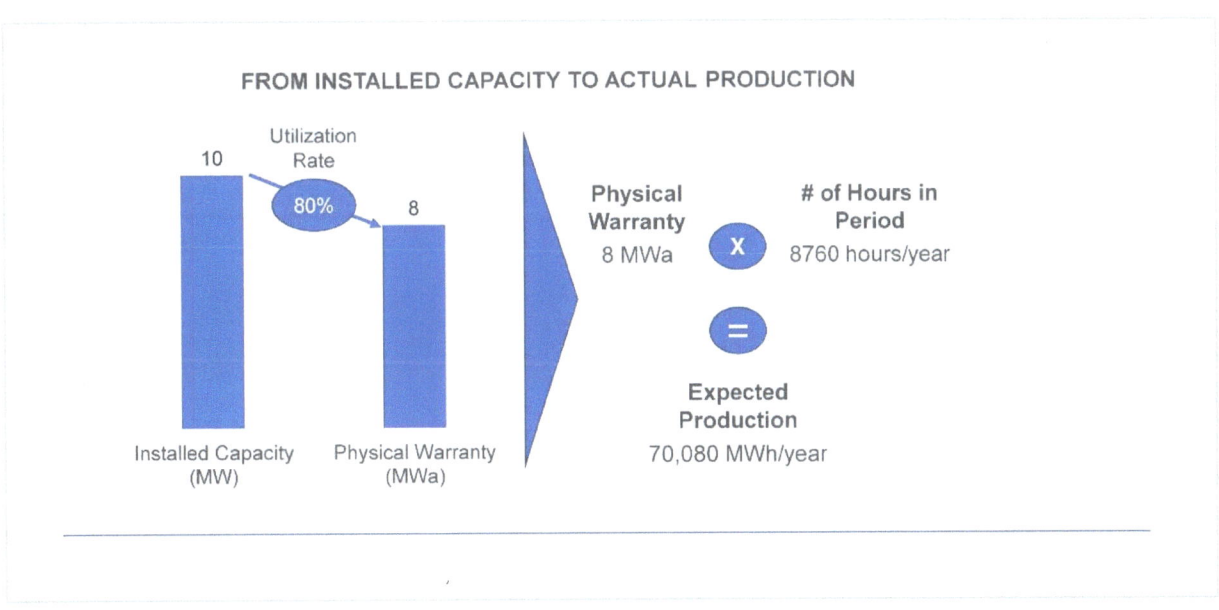

FROM INSTALLED CAPACITY TO ACTUAL PRODUCTION

Figure 4 - Overview of Relationship of Installed Capacity, Actual Production, Physical Warranty and Utilization Rate

4.4.2. ACL – "Environment for Free Purchase"

Representing 27% of the energy sold in Brazil, as shown in Graph 10, the ACL, or "Ambiente de Contratação Livre", allows free trade of electricity in the Brazilian market. Such trades are done bilaterally, with overseeing of the CCEE, and as opposed to the regulated environment (ACR), trading at the ACL does not occur through auctions.

There are two ways to purchase electricity in the ACL – either participating as a "Free Buyer" in this market, a position exclusive for consumers that have a minimum demand of 3 MW on average in each year[ix], or by joining as a "Special Buyer", a position that allows consumers with average demands between 0.5 and 3MW to purchase at the ACL exclusively from Small Hydropower Plants (PCHs). The types and quantity of players can be seen in Graph 11, which shows a total of 613 Free Buyers, 1,142 Special Buyers, and 584 Power Suppliers (Generators and

---

[ix] A 3 MW average demand in a given year effectively means 26,280 MWh consumed between January 1st and December 31st of that year. As the year has approximately 8760 hours, 26,280 Megawatt-hours consumed within a year mean an average demand of 3MWa through the year.

Independent Producers). Anyhow, consumers agree on purchasing a given amount of energy, in MWh, to be used over a pre-determined period of time. Unused energy is liquidated through the PLD, as explained ahead in section "4.4.3. Clearing the Market – CCEE's PLD".

Furthermore, Graph 12 serves to better illustrate the consumption proportion and the final usage of electricity from the ACR and the ACL. Because of the consumption size restrictions, over 92% of the energy traded in the ACL is utilized by the industry sector, as it can be better observed in the Graph 13.

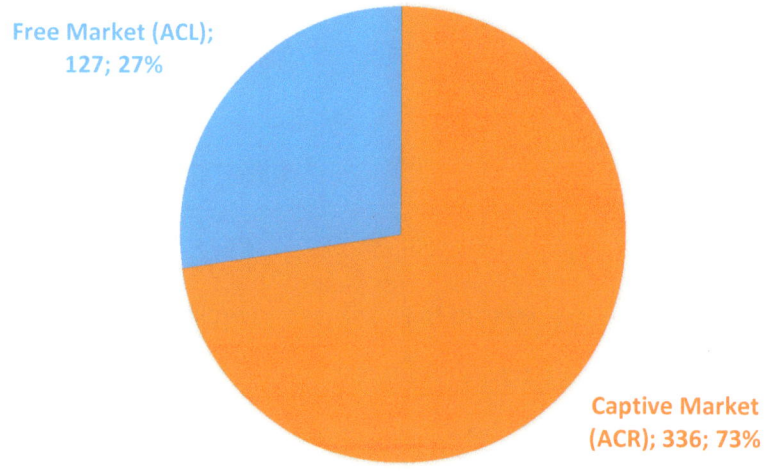

ELECTRICITY CONSUMED BY MARKET
Market; TWh; Share (2013 data)

Free Market (ACL); 127; 27%

Captive Market (ACR); 336; 73%

Source: Balanço Energético Nacional 2014 - EPE

Graph 10 - Electricity Consumed in Brazil by Market Type

TYPES AND NUMBER OF PLAYERS
Brazilian Electricity Market in Number of Players

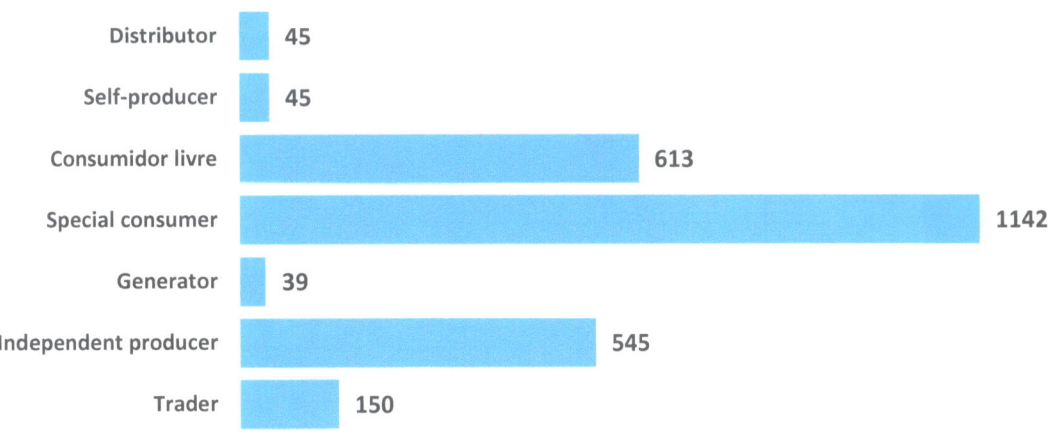

Source: Anuário Estatístico de Energia Elétrica 2014

Graph 11 - Type and Number of Players in the Brazilian Electricity Market

At the ACL, consumers purchase energy through bilateral agreements, and to the moment that this thesis was written, the conditions of such agreements were kept confidential and only disclosed at the will of the parties involved in each contract. Having said that, there is a strong lack of data available on prices, quantities and other measures related to the free trade of electricity. Through interviews with investors and other stakeholders, it was possible to have some reasonable idea of the current prices of energy, and of the length of the term of such bilateral agreements. With the caveat of no extensive data analysis, it has been reported in such interviews that such contracts are mostly anywhere from one month to five years in length. Most often, they are referred to as "short-term", usually from one to six months, or "long-term", often covering periods of 2 years or more. Due to their bilateral nature, such contracts could cover much longer terms, e.g. 10 years, depending on buyer and seller needs. Examples of this can be found in reports from sources such as investors' relations and news websites (Enfoque, 2005).

ELECTRICITY CONSUMPTION - BY CLASS

Consumption Class; TWh; Share

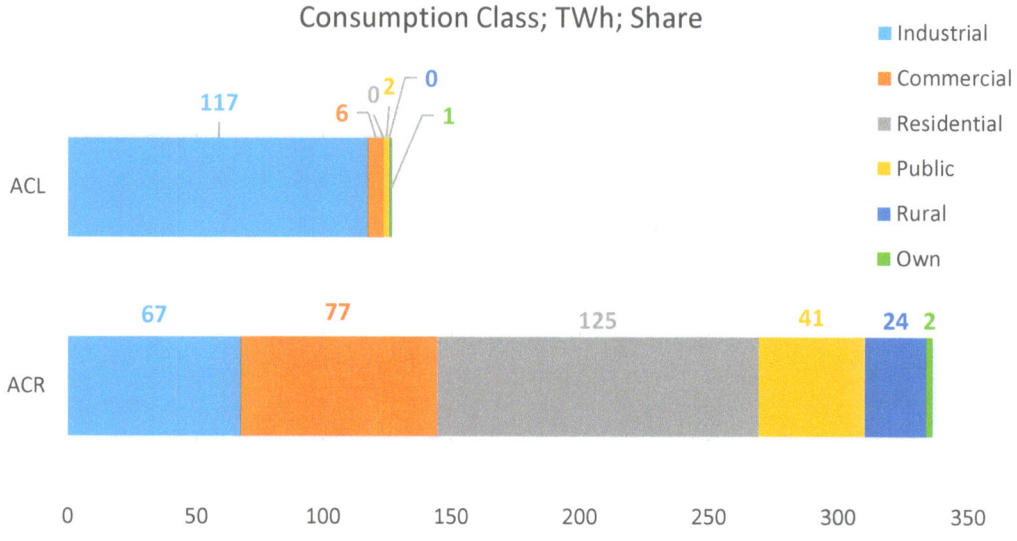

Source: Anuário Estatístico de Energia Elétrica 2014

Graph 12 - Electricity Consumption by Class

ELECTRICITY CONSUMED AT THE ACL

End Usage; TWh; Share (2013 data)

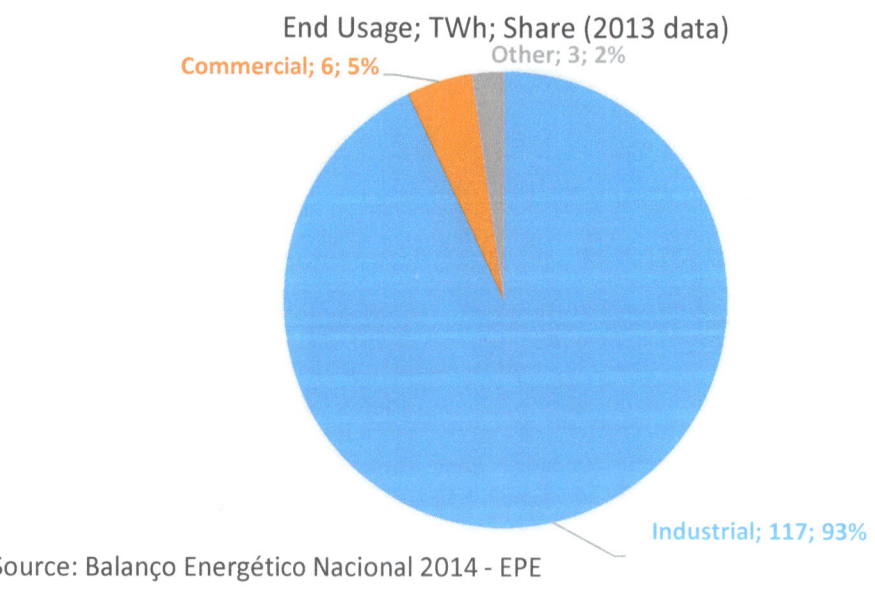

Source: Balanço Energético Nacional 2014 - EPE

Graph 13 - Electricity Consumption at the Free Market (ACL) by Class

4.4.3. Clearing the Market – CCEE's PLD

CCEE is the institution responsible for the "clearing" of the electricity market, and it does so through the PLD – the "Preço de Liquidação das Diferenças", or "Price of Liquidation of Differences". Due to the nature of power plants, especially the hydropower ones that strongly depend on weather, at the end of each month the CCEE "liquidates the differences" through the PLD. The system is rather simple: A plant may sell up to 100% of its Physical Warranty in advance, in the ACR or the ACL. After a given month is over, if such plant supplied less power than it previously sold, it is forced to "purchase" the power needed to complement its sales from plants that supplied more than sold in that same month. The price of such transaction is the same for all plants – it is the PLD price, officially defined weekly by the CCEE. The PLD is determined based on the marginal cost of generation, or in other words, the marginal cost for the most expensive plant needed to operate in that given week.

By definition, the lowest value possible of the PLD should be equivalent to the Marginal Cost of the MWh of the Hydropower Plant of Itaipu – about BRL 15/MWh. On the opposite extreme, the highest value should be defined by the marginal cost of the most expensive Thermal plant with capacity above 65MW, currently the plants of Termomanaus and Pau Ferro, both with marginal costs above BRL 1,100. Historically, in times of full production of hydropower (i.e. with healthy raining weather), the PLD has been often less than BRL 40/MWh. In times of droughts, the PLD peaked at values above BRL 800/MWh in nominal terms (such as in November of 2011), or at values above BRL 1,200/MWh in "February-2015 BRL" (real terms (IBGE, 1994-2015)[x], as it occurred in late 2001) (CCEE, 2001-2015)[xi]. Despite of the marginal cost behind it, since January of 2015, the government instituted a ceiling of BRL 388.04/MWh for the PLD, as well as a floor of BRL 30.26 (ANEEL, 2014). Graph 14 ahead shows in greater detail the behavior of the PLD over the 2001-2015 period.

---

[x] Real terms based on IPCA monthly historic data since 1994 can be found at IBGE: ftp://ftp.ibge.gov.br/Precos_Indices_de_Precos_ao_Consumidor/IPCA/Serie_Historica/ipca_201502SerieHist.zip
[xi] Historic PLD prices may be downloaded in CSV format at CCEE's website, at http://www.ccee.org.br/portal/faces/pages_publico/o-que-fazemos/como_ccee_atua/precos/precos_csv

Lastly, although outside the scope of this thesis, it is important to mention that these limitations to the PLD imposed by the government do not impact the true costs of the systems. It could, in fact, arguably increase the long-term prices of energy due to the significantly smaller financial signal sent to consumers and suppliers exposed to the PLD, as defended by the Brazilian Association of Electricity Traders (ABRACEEL, 2014)[xii]. This thesis does not argue against or in favor of such argument.

The PLD plays an important role for the ACL: not only it defines the "spot price" for the free market, but also it is an important variable in the definition of longer term prices of electricity. Based on the PLD, institutions like the private company BBCE (BBCE, 2015) dedicate significant effort in creating indexes (such as BBCE's "Índice Spot") that allows companies to trade energy in the short term. Long-term ACL prices are also influenced by the PLD – when the PLD is on its low-end, about BRL 30-40/MWh, the consumer's cost of waiting for a better deal is not only drastically reduced, but in fact often negative. In this situation, consumers buy on the long-term to reduce the risks of price variation (i.e. "better to pay BRL 100/MWh during 5 years than pay BRL 30 today and possibly BRL 388 next month"). On the other hand, when PLD are on their high-end, like it has been since early 2014 to the writing of this thesis, the option of waiting for a better long-term deal becomes very costly for consumers, and selling long-term only becomes interesting for power suppliers if done at a much higher price than during the PLD's low-end.

As described above on the ACL details, there is no publicly available long-term free market data for electricity prices in Brazil. Thus, understanding with an in-depth level of detail how the length of the contract's term and the current and past PLD affect the current ACL prices is merely a qualitative exercise. With the data acquired through researching and interviewing key stakeholders for this thesis, the work developed here leads to an expectation of a price of BRL 100-150 at PLD bottoms, and of BRL 220-280 for 2-5 year ACL contracts. However, there is insufficient data to have an acceptable level of certainty over such historical prices. Lastly, there is a judicial fight over a law that would make mandatory the disclosure of prices of bilateral

---

[xii] More on this topic and ABRACEEL's position can be found at the following link, retrieved March 13th, 2015: http://www.aneel.gov.br/aplicacoes/consulta_publica/documentos/Abraceel%20CP%20009_2014.pdf

agreements of energy trade (ACL), but so far the Brazilian justice has detained CCEE from publicizing such data.

PLD HISTORIC PRICES 2001-2015

PLD prices per MWh, Real terms (BRL Feb-2015, inflation by IPCA)

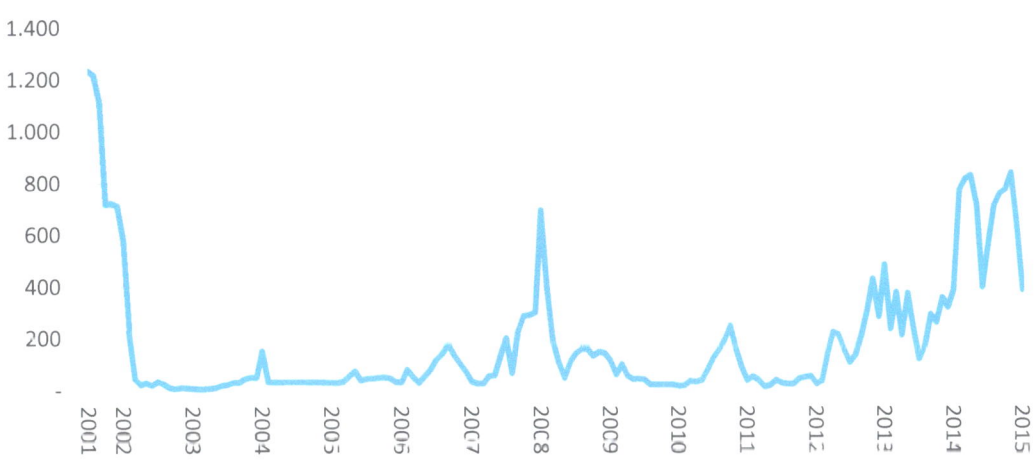

Source: CCEE - Resultado Consolidado de Leilões - February/2015

Graph 14 - Historic Prices of PLD for the 2001-2015 Period in Real Terms

4.4.4.   The MRE – "Mechanism for Energy Relocation"

The MRE, or "Mecanismo de Realocação de Energia", is an electricity pool created by the CCEE that allows plants to reduce their exposure to the PLD variation. The first objective behind the conception of the MRE was to share the financial risks associated with the energy trading done by hydropower plants that were centrally dispatched by the National System Operator (ONS). The nature of hydropower has an inherent risk of production – variations on precipitation can deeply impact hydropower generation, as it is described in detail in the section "4.5. Expansion plan, Economy Growth and the Drought" of this thesis. Furthermore, the ONS dispatches plants based on optimization of resources, thus trying to reduce the overall cost of the system and not of a single plant. This sometimes means draining faster reservoir A than B, which could create a somehow unfair and risky operation for hydropower investors. Moreover, many hydropower plants are built cascaded on the same river, and thus the good operation of one may cause bad results for the next one downstream.

This intrinsic impossibility of precisely forecasting the production of hydropower plants also creates a financial challenge. On one hand, suppliers and consumers need to be able to negotiate electricity in advance, while on the other hand the uncontrollable risk of hydropower operation would not allow suppliers to do so without exposition to the volatile PLD. The MRE then comes into play – it allows hydropower generators connected to any point of the SIN (the national's electricity grid) to join a pool of generators that agrees to sell excess or buy shortage of energy at a very low price. The MRE's price works like a "PLD substitute", and is known as TEO – "Tarifa de Energia de Otimização" or "Tariff of Energy Optimization". The TEO is fixed annually by ANEEL, and its current value, for 2015, is fixed at BRL 11.25, or 2.9% of February's PLD (ANEEL, 2014). The definition of the TEO is comprised by the sum of the lowest Operations and Maintenance cost and the "CFURH", an acronym that stands for the "Financial Compensation of the Utilization of Hydric Resources". Thus, by definition, the TEO cannot raise substantially year-over-year, meaning a significantly reduced exposure for suppliers at the MRE, even in the long-term.

For Small Hydropower Plants, the participation in the MRE is facultative. Depending on the portfolio of sales of a PCH, the trade-offs of joining or not the MRE change significantly. For instance, to a PCH that sold 100% of its Physical Warranty on an A-3 auction, for a period of 30 years, joining the MRE reduces the standard deviation of the present value of its energy sales by an order of magnitude. A numerical example serves well to enlighten this topic: in 2014, PCHs with such portfolio sold energy for about BRL 165/MWh. Joining the MRE means that for every extra MWh the plant produces, it will receive the TEO (BRL 11.25, or ~7% of the agreed price, slightly readjusted annually), while for every MWh it lacks in production it will pay the same TEO for another supplier to cover its client's contracted purchase. If not joining the MRE, every MWh it lacks may cost the PLD – anywhere between BRL 30.26 and 388.04, or ~18% to 235% of the agreed price, abruptly readjusted monthly – while it will benefit by selling extra production for the same PLD.

As data shows, a few months of production significantly lower than the Physical Warranty may cause the investors of a PCH to go bankrupt. In fact, the large financial risk of PLD exposure was one of the reasons for the cap of BRL 388 instituted in 2015. As AES Brasil, one of the largest energy companies in the country stated, "The elevated PLD may generate a financial pressure on distributors, Free Buyers and generators, possibly creating unsustainable businesses and leading companies to bankruptcy" (PortalPCH, 2014). Lastly, CCEE's January report "InfoMercado 2015" shows the importance of the MRE for PCHs – of the 434 PCHs operating, 75% chose to be part of the MRE, as shown in Graph 15.

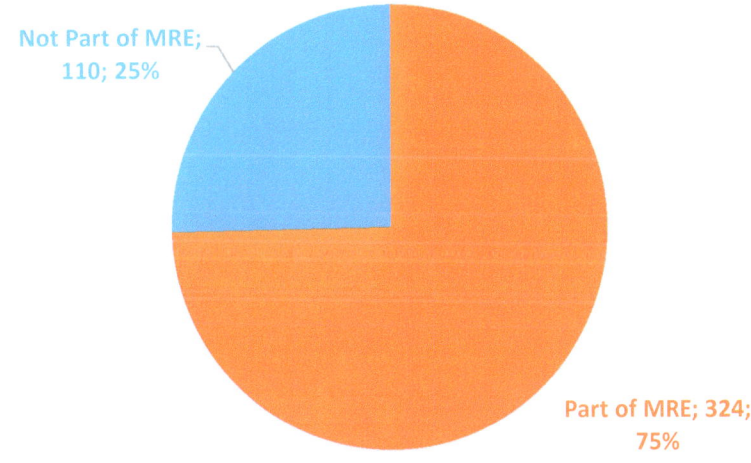

PARTICIPATION ON THE MRE BY PCHS

MRE Choice; Number of PCHs; Share

Not Part of MRE; 110; 25%

Part of MRE; 324; 75%

Source: CCEE InfoMercado 2015

Graph 15 - Participation on the MRE by PCHs in 2015

4.4.4.1.   Expulsion of the MRE

ANEEL and the CCEE regulate the MRE in a way that ensure that the players within the pool are performing well over time. In other words, the pool protects players during difficult times, but it excludes players that present a consistently low performance. The performance of each plant in measured by comparing the average annual generation of its lifetime to its Physical Warranty. The Physical Warranty is the expected annual generation of a given plant, and it is defined by ANEEL based on the physical characteristics and past performance of the plant. As explained in section "4.4.1. ACR – "Environment for Regulated Purchase" and shown in Figure 4, it is measured in Average Megawatts of output per year, i.e. 1 MWa of Physical Warranty means that the plant is expected to produce 8760 MWh per year of energy.

The Table 1 below shows the thresholds established by the Normative Resolution Number 409, from August 10th, 2010 (ANEEL, 2010). Such thresholds are defined to take into account that newer plants may present low average performance due to the small number of data points available and one-time episodes that might have driven the average lower than expected. On the other hand, plants that have been operating for longer periods are required to present much higher averages, as such one-time events would have been diluted over the large number of data points utilized. As an example, for a plant operating for more than 10 years, the minimum average performance required to remain within the MRE is 85%. This means that if the average annual generation of the plant, considering the period since the beginning of its operations, is lower than 85% of its Physical Warranty, the plant should be excluded of the MRE.

Table 1 - Thresholds for exclusion of the MRE, as in Resolution number 409 of ANEEL

Number of months registered at CCEE after the 12th month of commercial operation (m)	Minimum Average Generation/Physical Warranty (in %) Required
24 ≤ m < 36	10%
36 ≤ m < 48	55%
48 ≤ m < 60	60%
60 ≤ m < 72	65%
72 ≤ m < 84	70%
84 ≤ m < 96	75%
96 ≤ m < 120	80%
120 ≤ m	85%

4.4.5. Self-Production – the Role of the "Auto-Produtor"

In 2013, there were 45 companies categorized as Self-Producers, owning a total of 16% of the installed capacity of electricity generation in Brazil. Such companies are mostly industrial players in sectors where the cost of electricity is a relevant part of their operations – often meaning 15 to 45% of their costs, depending on the sector. Those companies not only benefit from the control over the supply of one of its most important cost-drivers, but also they gain with government subsidy, given to consumers that also own generation.

This thesis will not focus on the trade-offs of self-production, but Small Hydropower plants tend to be very attractive to self-producers due to its small size, reduced environmental impact, small risk and easiness of execution. Such self-producing consumers benefit from the subsidy in form of exemption of some governmental fees (Proinfa, the fee used for incentivizing clean sources of energy; CCC, a charge based on the consumption of combustibles; the CDE, a fee used to sponsor development of the electricity sector in certain states), the exemption of the TUSD, which is the tariff of distribution of electricity, and a reduction in overall taxation. The largest sectors in self-production are Steel, Aluminum, Pulp and Paper, Cement, Chemicals and Sugar and Ethanol.

Lastly, the self-production share in installed capacity has been growing expressively in Brazil, from levels of 6% in 2001 to 16% in 2013, the latter shown in Graph 16. Within self-production, Hydropower has also been growing, from 21% of the total in 2001, when thermal power represented 79%, to 24% in 2013, when thermal represented the other 76% (EPE, 2014).

INSTALLED GENERATION CAPACITY - BY FOCUS
Focus (market or self-production); GW; Share

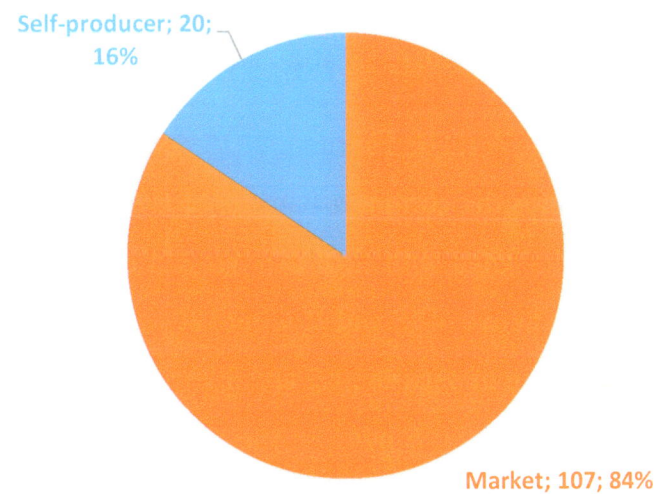

Source: Balanço Energético Nacional 2014 - EPE

Graph 16 - Installed Generation Capacity by Focus of Utilization

41

4.5. Expansion Plan, Economy Growth and the Drought

4.5.1. Current Plan: Power Plants and Expected Entry Time

In 2007, the Brazilian government issued the "Plano Nacional de Energia 2030" (EPE, 2007), which together with the document "Matriz Energética Brasileira 2030" released in November of 2007, define the long-term strategy of the Energy supply of the country. This planning was made through the MME, in partnership with the EPE and the CEPEL ("Centro de Estudos e Pesquisas em Energia Elétrica"). This plan covers about 20 years of development, mainly due to the long-term required for significant changes of expansion of the energy supply and demand. As a quick reference, building a single large hydropower plant takes about 4-5 years.

The PEN 2030 plan was made during a very prosperous time for Brazil – in 2007, the GDP grew by 6.1% in real terms, coming from a compound annual growth rate of 4.7% on the 2004-2007 period (IBGE, 2015). As it can be seen on the PEN 2030's table 4.8, of their 4 scenarios, the best-case expected the economy to grow 5.1% per annum during the 2005-2030 period, while the worst-case expected the country to grow at 2.2% per annum in the same period.

In more recent days, the ANEEL's February/2015 document "Resumo Geral das Usinas" (ANEEL, 2015), which evaluates the current expansion of electricity supply under construction, shows a supply growth well aligned with such GDP growth expectations of the PEN2030. The electrical supply will increase between 18 and 25% until 2018, as shown in Graph 17, thus having a compound annual growth rate of between 4.2 and 5.7%. This document also shows almost no increase in supply after 2018, but this is because the construction time of plants is rarely more than 4-5 years, thus there should be virtually no plants being built today that will only enter the market after 2019.

Considering the absolute installed capacity entering the Brazilian market each year there is another indication of the strong supply growth the country should face in the next 5 years. The period 2001-2014 had an average annual addition of 4.2GW, meaning an average increase of 4.1% in the installed capacity each year. The same metrics for 2015-2018 are expected to be 8.2GW and 5.3% respectively, as shown in Graph 18 below.

INSTALLED CAPACITY AND UNDER CONSTRUCTION

Installed Capacity in GW and forecast based on current constructions

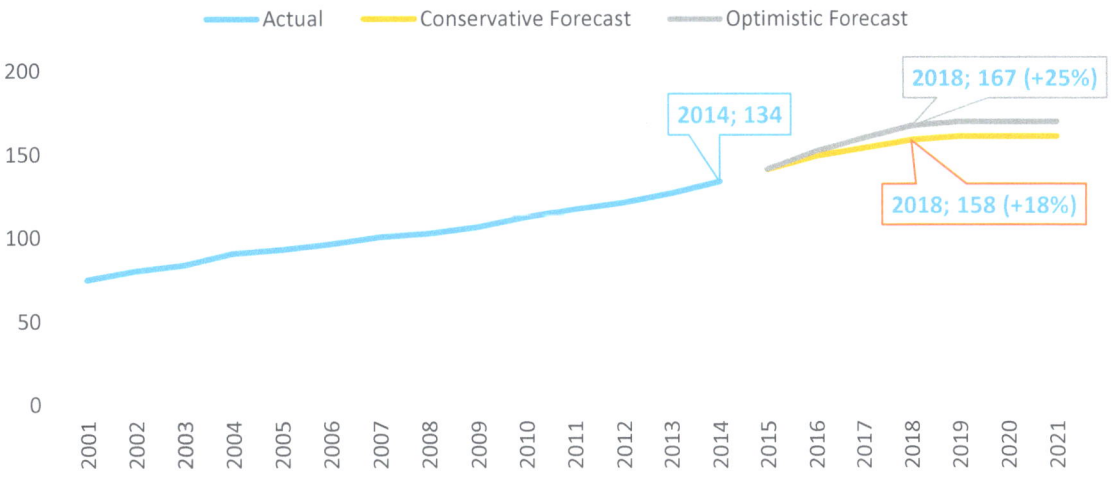

Source: ANEEL - Resumo Geral das Usinas - Feb/2015

Graph 17 - Installed Capacity and Capacity Under Construction

ADDED CAPACITY PER ANNUM

Installed Capacity in MW and forecast based on current constructions

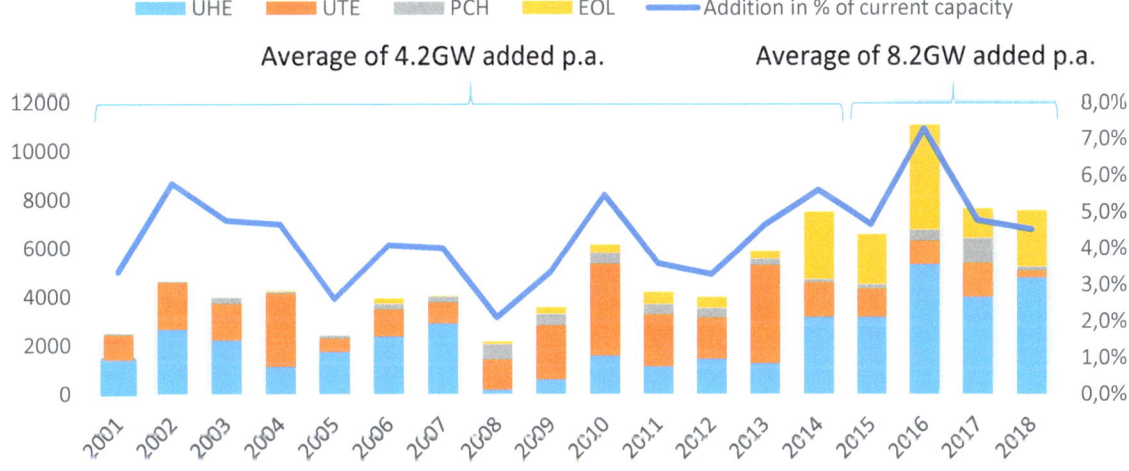

Source: ANEEL - Resumo Geral das Usinas - Feb/2015

Graph 18 - Annual Addition of Capacity for the 2001-2018 Period

4.5.2. Lack of Economic Growth in Comparison to the PEN2030

The Brazilian GDP growth rates in 2008 and 2010 were great – 5.2 and 7.5% respectively. Even when not taking into account the recession in 2009 (-0.3%), Graph 19 shows that PEN2030's (EPE, 2007) perspective for Brazilian GDP assumed far more growth than the recent reality. Furthermore, the IMF 2015-made forecasts expect the country's GDP to continue growing at a reduced rate (IMF, 2015), close to the worst-case scenario. Graph 20 also provides a quick comparison of the accumulated growth expected for this decade (2011-2019 period), and shows that the current IMF perspective is about 5% smaller in such metric than the PEN2030's worst-case scenario.

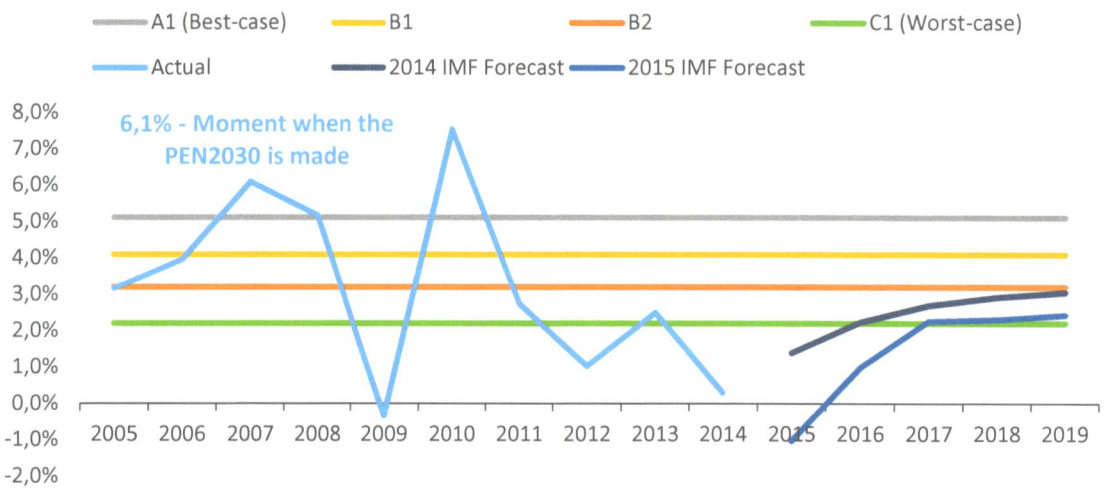

Source: IBGE; IMF World Economic Outlook; PEN 2030

Graph 19 - GDP Growth - Historic and Current IMF Forecasts versus the PEN 2030 Scenarios

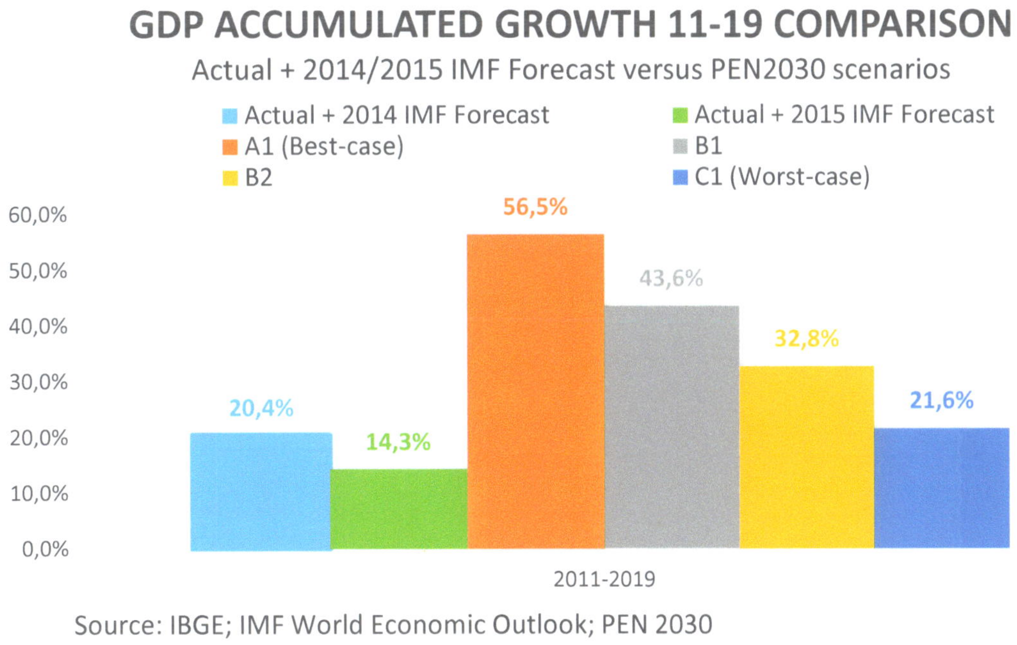

GDP ACCUMULATED GROWTH 11-19 COMPARISON
Actual + 2014/2015 IMF Forecast versus PEN2030 scenarios

Legend:
- Actual + 2014 IMF Forecast
- Actual + 2015 IMF Forecast
- A1 (Best-case)
- B1
- B2
- C1 (Worst-case)

Chart data (2011-2019):
- 20,4%
- 14,3%
- 56,5%
- 43,6%
- 32,8%
- 21,6%

Source: IBGE; IMF World Economic Outlook; PEN 2030

Graph 20 - GDP Accumulated Growth for the 2011-2019 Period

4.5.3. The Effects of the GDP in the ACL

As seen previously in this thesis, more specifically in Graphs 12 and 13, the ACL is composed 93% by industrial consumption. Moreover, 64% of the industrial consumption of electricity is done within the ACL (EPE, 2014)[xiii]. Thus, the demand within this market is highly dependent on the industrial consumption of electricity in the country. On a different perspective, as expected, data analysis of the 2000-2013 period show a significant correlation ($R^2 = 0.6053$) between the variations of GDP and industrial electricity consumption, as can be seen in the Graph 21. Thus, an investor planning to supply electricity to the ACL should probably take into account the expectations for the GDP when understanding the demand side of the market.

Furthermore, the uncertainty implied in the GDP variations, inclusive the ones demonstrated by the difference in the PEN2030 and the actual data, should always be taken into account by investors. One implicit way this takes effect is through the very high rate of returns required in Small Hydropower Plants projects. Even seasoned investors that have reduced capital

---

[xiii] 117TWh out of the 185TWh total industrial consumption of electricity in 2013, according to the "Anuário Estatístico de Energia Elétrica 2014"of the EPE

45

expenditure risks and have already walked the learning curve still demand very high rates of return, not unusually on the 20-30% per annum range.

GDP GROWTH AND INDUSTRIAL CONSUMPTION
GDP growth (X) versus Industrial electricity consumption growth (Y)

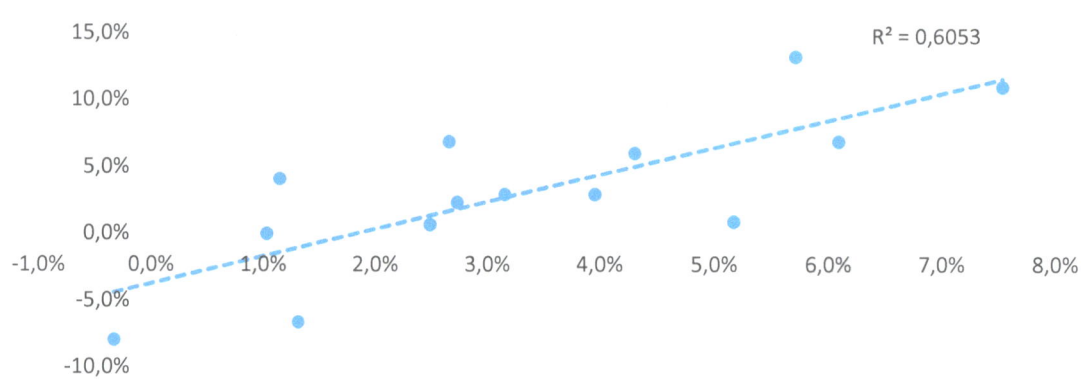

Source: IBGE; IMF; PEN 2030

Graph 21 - GDP Growth and Industrial Consumption Growth Comparison

As a simple exercise to stimulate the imagination of the reader, the same equation of the linear correlation between GDP growth and Industrial Electricity Consumption found in the 2000-2013 period was applied for the IMF forecasts until 2019. The results are displayed in Graphs 22 and 23 for IMF forecasts made in 2014 and 2015 respectively. They show a growth much slower than what was forecasted at the PEN2030 and what can be seen in the previous section "4.5.1. Current Plan: Power Plants and Expected Entry Time".

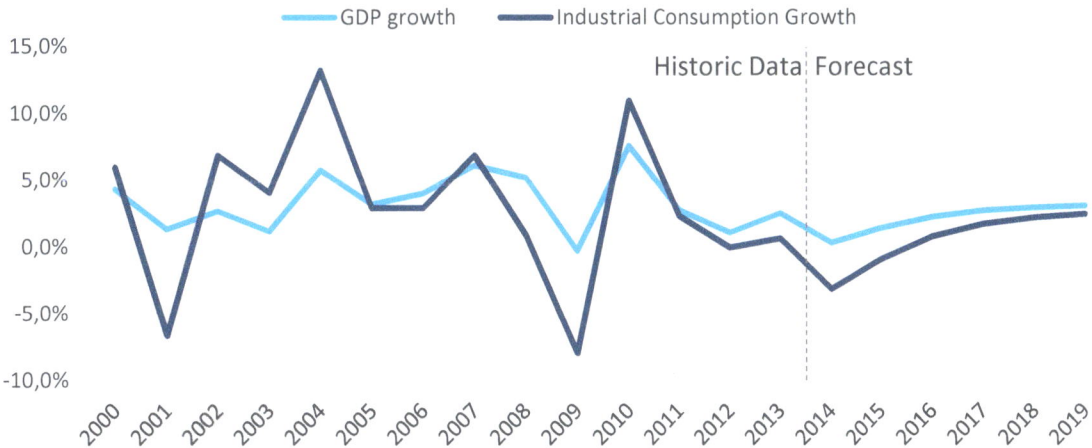

Graph 22 - Industrial Consumption Estimative Based on IMF 2014 GDP Growth Forecasts

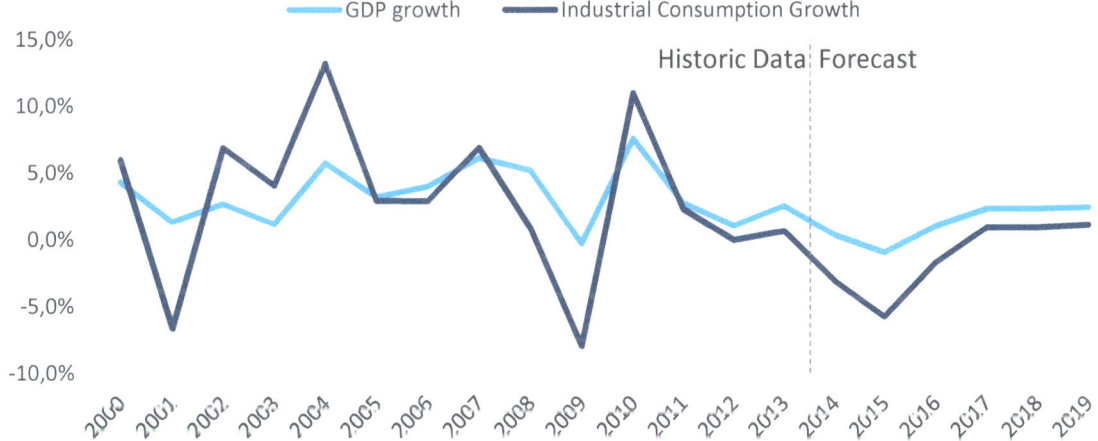

Graph 23 - Industrial Consumption Estimative Based on IMF 2015 GDP Growth Forecasts

4.5.4. The Current Drought – Unequaled in the Last 60 Years

To understand the past behavior of the weather is already a complicated matter – even more so for an electrical engineer or a business administrator. Moreover, to forecast it with precision over long periods of time is an order of magnitude more difficult than that. However, the impact of the precipitation on hydropower generation is clear – and within a country that has 70% of its installed capacity based such generation, it becomes even more evident. According to CCEE's InfoMercado detailed data on generation, the 170 operating large hydropower plants generated only 80% of their Physical Warranty – meaning that there was a lack of around 10.1GW of average generation during the year. This is approximately equivalent of 9.5% of the country's installed capacity being offline[xiv] during 2014.

For Small Hydropower, the situation is a little better – not to mention that, by nature, the impact in the country is considerably smaller. In 2014, the PCH's produced only 85% of what was expected – or 0.4GW of average generation lacking during the year. However, a grim outlook was presented to the PCHs exposed to the PLD. As a quick way of measuring this impact in financial terms, if during 2014 all PCHs would not be part of the MRE and would be exposed to the PLD in its current price – BRL 388.04/MWh – this would mean losses of approximately BRL 1.37 Billion[xv] for PCH investors.

4.5.5. The Current Drought – Understanding Precipitation

As mentioned earlier in this thesis, during the interviews performed investors were confident that the drought would not allow hydropower generation in the country to get back to normal levels within the next 3-5 years. To better understand the behavior of precipitation in the Brazilian territory over the past years, valuable data was collected at the database of INMET, the Brazilian National Institute of Meteorology. The institute had data on 593 weather stations spread in the Brazilian territory, as shown in the map of Figure 5. Out of those, 122 stations had

---

[xiv] Assuming a factor 80%, 10.1GWaverage would mean 12.6 GW in installed capacity, or approximately 9.5% of the 2014's 133GW of installed capacity in Brazil.

[xv] PCHs produced 20,144 GWh in 2014, while they should have produced 23,673 GWh. If the 3,529 GWh missing would have to be repurchased by the PCHs at the PLD of BRL 388.04/MWh, this would mean BRL 1,369 Million.

monthly data for all months between January 2001 and 2015, and thus their total was utilized as a proxy for precipitation in Brazil, as shown in Graph 24. This results in a very superficial analysis, with the single objective of developing some intuition on the variation of precipitation and its accumulation over time.

PRECIPITATION IN THE 2001-2015 PERIOD

Historic monthly data of precipitation, relative to Jan-01

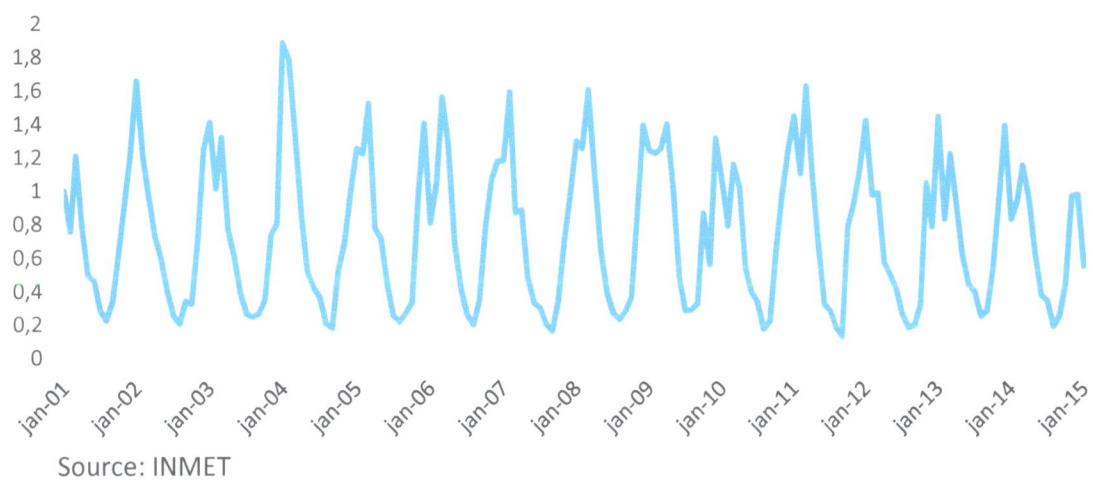

Source: INMET

Graph 24 - Monthly Precipitation Data for the 2001-2015 Period

Figure 5 - Map of INMET precipitation data collection stations

Looking at Graph 24, it is possible to tell by inspection that, in the more recent years, precipitation has been on lower levels than in the 2001-2005 period. However, the hydropower generation does not vary rapidly with monthly precipitation rates. Instead, it varies much more correlated with the long-term accumulation of rain. The vast majority of hydropower in Brazil comes from large plants, and those always have large reservoirs acting as "water buffers", enabling them to maintain healthy levels of generation. The largest reservoir in Brazil is located in Sobradinho, BA, and has 4,214km^2 of inundated area. Itaipu, the largest hydropower plant in Brazil, has a reservoir that covers 1,350km2 of area, or approximately the size of Phoenix, Arizona. To better understand this long-term variation, Graph 25 shows the 24-month accumulated precipitation since January 2003 (thus referring to the 2001-2003 period) until January 2015. Graph 26, on the other hand, shows the sum of actual water flow in the points of hydropower operation in country. Although it only covers until December 2013, it starts back in

1931 – thus showing how large droughts, such as the one from 1948-1956, may affect the generation market. As a consequence of such water flows, Graph 27 shows the actual percentage of the hydroelectric reservoirs capacity that was filled with water during the 2000-2015 period.

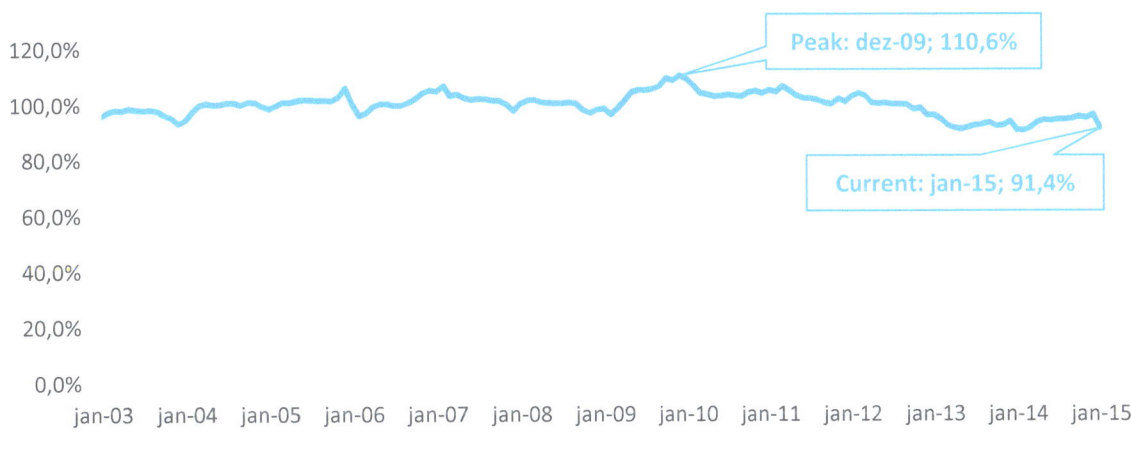

PRECIPITATION IN THE 2001-2015 PERIOD
Historic 24-month accumulated precipitation relative to period average (Average = 100%)

Source: INMET

Graph 25 - 24-month Accumulated Precipitation Data for the 2001-2015 Period

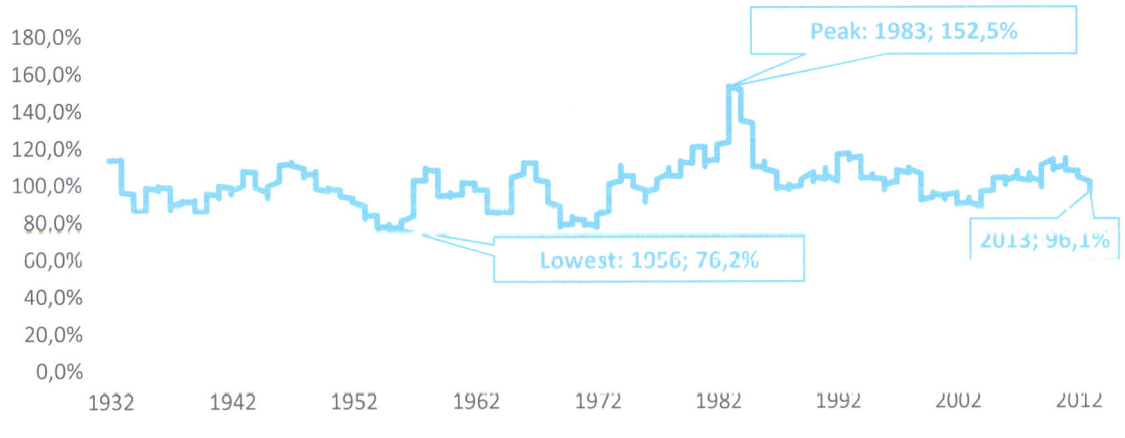

WATER FLOW IN THE 1931-2013 PERIOD
Historic 24-month accumulated waterflow in the points of operating hydropower in Brazil, relative to period average (100%)

Source: ONS

Graph 26 – Water Flow Data for the 1931-2013 Period

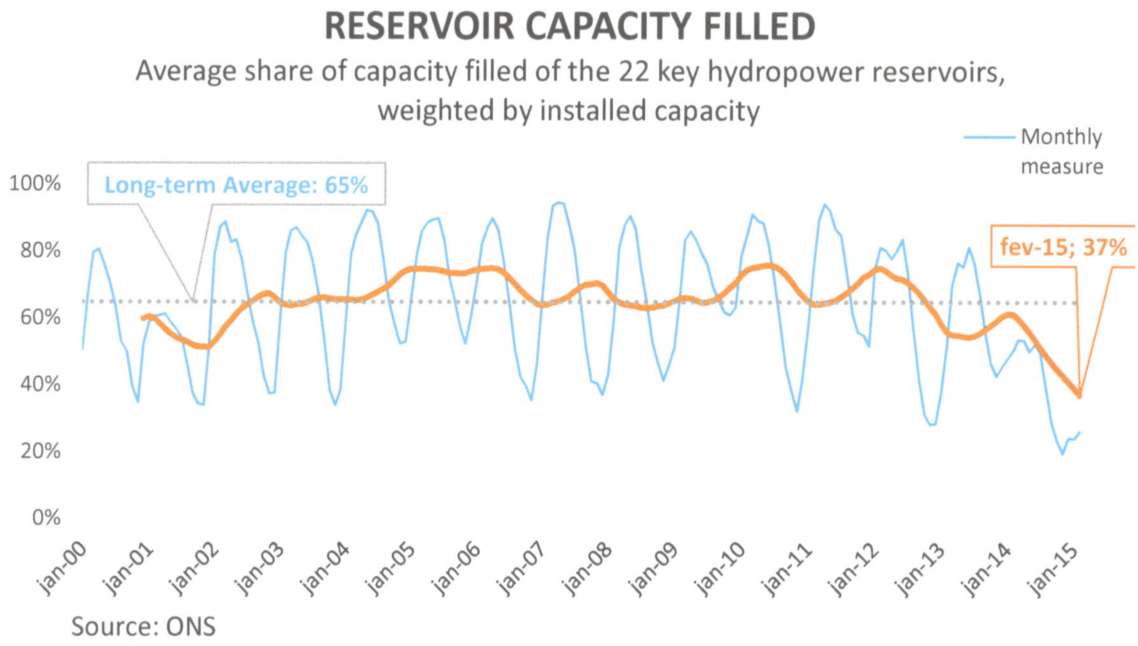

RESERVOIR CAPACITY FILLED

Average share of capacity filled of the 22 key hydropower reservoirs,
weighted by installed capacity

Graph 27 - Reservoir Capacity Filled Data for the 2000-2015 Period [xvi]

Lastly, it is also interesting to observe how quickly the accumulated precipitation and water flows in the hydropower system may recover to average levels. One example can be seen in the period after 1956: In February 1956, the hydro system had its flows at about 76.2% the historical average. In September of 1957, 19 months later, it was already back to the historical average (ONS, 2015). In fact, there are already small indications of a potential recovery of the water flows in the Southeast. One of the most important water reservoirs, the Cantareira, reported having "the most rainy February [2015] in the last 20 years, with precipitation at 61.9% above its historic average for the month" (G1 News, 2015).

Most importantly, the large variation in relatively short periods of time (1-2 years), shows that investors that intend to build Small Hydropower plants, which may take up to 2-3 years,

---

[xvi] The 22 key reservoirs on the SIN are: Furnas, Mascarenhas de Moraes, Marimbondo, Água Vermelha, Emborcação, Nova Ponte, Itumbiara, São Simão, Ilha Solteira, Barra Bonita, Promissão, Três Irmãos, Jurumirim, Chavantes, Capivara, Governador Bento Munhoz, Salto Santiago, Três Marias, Sobradinho, Luiz Gonzaga, Serra da Mesa and Tucuruí

should always be careful with optimistic calculations. In other words, a drought like the current one may take spot prices to very interesting levels for investors, but a situation like this may not last long enough to provide appropriate returns to those that choose to remain exposed to the PLD.

Lastly, it is important to understand how erratic and unpredictable the precipitation process is. Even the very sophisticated Climate Prediction Center, of the National Weather Service of the United States Government, only provides predictions for up to three months (National Weather Service, 2015). Moreover, the impact of precipitation on hydropower generation is buffered by the reservoirs of the large plants. Since those represent almost 70% of the Brazilian installed capacity, and since the relationships between precipitation and reservoir accumulation is not exactly clear, it is of vital importance to explore the past behavior of the reservoir themselves, as well as their correlations with the electricity price. This discussion continues on the following section.

4.5.6. The Impact of the Drought on Electricity Prices

As already noted, the drought reduces the water flow in the Brazilian hydric system, which then causes reservoirs of hydropower plants to have much lower volumes of water. Graph 28 shows the impact of lower reservoir levels on the actual production of hydro. To eliminate the impact of the growth in installed capacity, this chart compares the utilization rate of hydropower, i.e. the actual production divided by the installed capacity at each given year, with the share of volumetric capacity filled with water on the 22 key Brazilian reservoirs.

With a significant reduction in production of hydropower-based electricity, which represent about 70% of the country's installed capacity, there is a shortage in low-marginal-cost supply. Since the PLD is defined by the marginal cost of the most expensive power plant required to "liquidate the differences", a reduction in hydropower means a more likely utilization of thermal-based plants for that end. Such plants have much higher marginal costs, and thus cause significant spikes in the PLD. As explained before in this document, the PLD is capped at BRL 388.04 since January 2015, but the period 2001-2014 in Graph 29 demonstrates the relation between the PLD and the reservoir levels in a very direct, temporal way.

Lastly, as explained in the section "Clearing the market – CCEE's PLD", the levels of the PLD directly affect the value of the option of short-term exposure versus long-term contracts, causing the bilateral ACL prices to fluctuate highly correlated to the PLD.

RESERVOIR VOLUME VS HYDRO PRODUCTION

Hydropower utilization rate (Generation/Installed Capacity) Vs. Weighted average share of key reservoirs filled

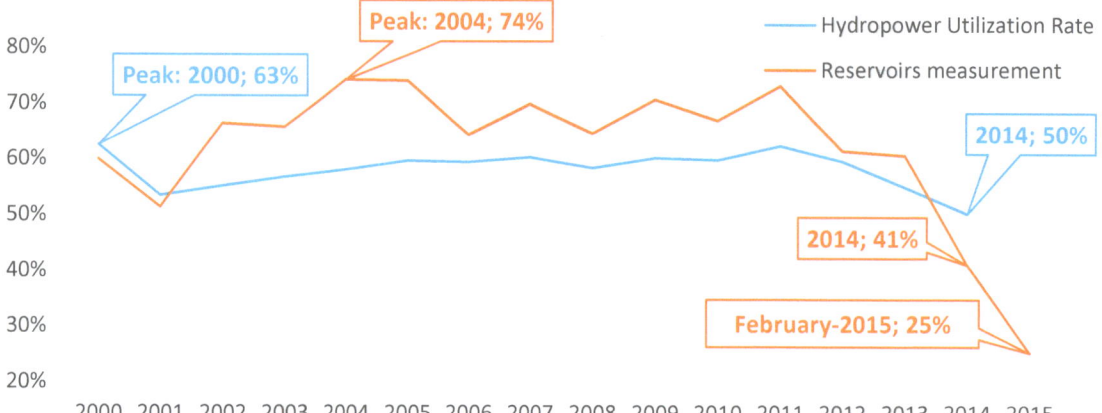

Source: ONS; CCEE; BEN 2014

Graph 28 - Reservoir Volume versus Hydropower Production for the 2000-2015 Period

RESERVOIR CAPACITY FILLED VS. PLD

Weighted average share of capacity filled of 22 key reservoirs vs. PLD

Source: ONS; CCEE

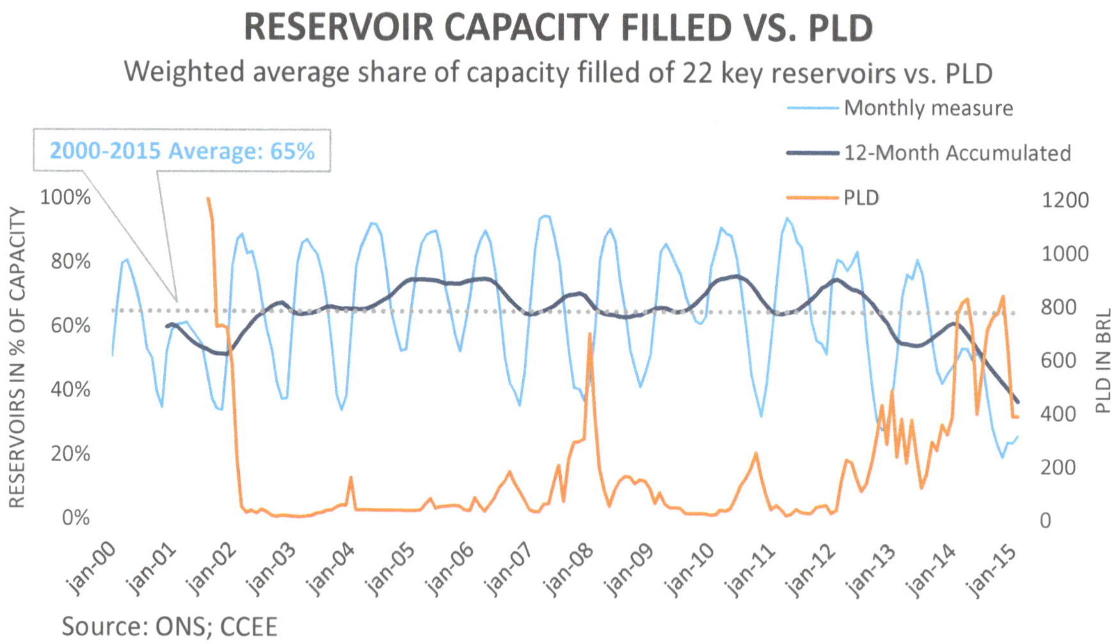

Graph 29 - Reservoir Capacity Filled Versus the PLD in Real Terms for the 2000-2015 Period

4.5.7. Great Short-Term, Uncertain Long-Term

Putting together the insights reviewed in this chapter so far inevitably leads the reader to question the long-term return-on-investment of a power plant in Brazil. It is, indeed, a market facing a struggling supply, which reflects on high spot prices of electricity. PLD prices have been peaking for more than a year, providing great short-term returns to investors exposed to this month-to-month market. However, an investment in a Small Hydropower plant requires up to 2-3 years of investments and will bring revenues over a period of a minimum of 30 years. For this reason, this thesis focuses on understanding the long-term trends of the market as well.

As previously noted, the Brazilian electricity supply has been planned on top of very optimistic economic development scenarios, and recent years have been pointing towards a 2010s decade well short of GDP growth expectations. More specifically, industrial electrical consumption is expected to follow this trend and grow at a much slower pace than previously forecasted, deeply impacting the ACL demand. Adding to such sluggish demand growth, the current drought has disabled a large part of the Brazilian hydropower potential. Although this made spot prices very interesting to investors, it also means that a recovery in the water flows and reservoirs could rapidly increase supply of hydropower back to normal levels. All these variables together inevitably point towards a trend of having the PLD prices going back to normal levels, in a period that could vary anywhere from a few months to the 3-5 years expected by investors.

Due to such wide variety of risks, ranging from lack of economic growth to lack of precipitation, it is advisable that small hydropower investors should sell electricity by building a portfolio of long- and short-term contracts, balancing the capture of healthy revenues from high spot prices with the certainty of revenues from long-term sales. The next part of this thesis will discuss a few possible portfolio choices. The objective of such study is to create a good intuition of portfolio impact in the net present value of Small Hydropower investments, as well as to demonstrate a method of value comparison.

4.5.8. Hydropower Potential in Brazil

During the interviews performed, a common commentary was regarding the limits of hydropower expansion in Brazil. Indeed, there is a limit to the electrical power that can be extracted from water flows in any given region, and Brazil is not an exception. However, the hydropower potential of the country has been deeply studied, and an equivalent of 109 GW of generation capacity have already been inventoried. The inventoried capacity, according to EPE, "includes operating and in construction power plants and those for which a basic and feasibility study has been prepared" (EPE, 2014).

Within the inventory, 86 GW have already been captured and are part of operating hydropower plants. Out of the other 23GW, the latest ANEEL report on the entry of new power plants on the market (ANEEL, 2015) shows 18 GW and 2GW of large hydropower plants and PCHs respectively being built and entering the market until 2018. As future potential, although not studied in-depth, the government already mapped another estimated 27GW of hydropower potential. Lastly, this total of 136GW is not expected to be 100% of the hydropower capacity of the country. As the time series in Graph 30 shows, this total may expand while the government further studies the Brazilian hydric potential and as hydropower technologies improve.

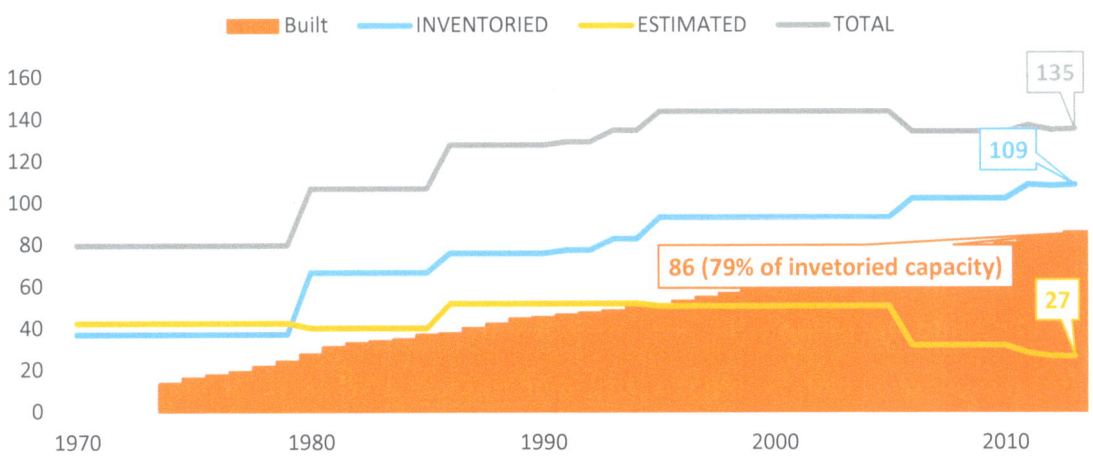

HYDROPOWER POTENTIAL
Hydropower Potential in GW of capacity

Source: Balanço Energético Nacional 2014 - EPE

Graph 30 - Hydropower Potential in Brazil

4.6.   Observations in Investment Behavior

It is not unusual to see several reports, studies and dedicated analysts working hard on the science of forecasting – especially when it comes to the price of commodities. Electrical Energy is not different, even more so in liberal markets like Brazil. Different from many other commodities, electricity is constrained by the cabled connections it requires for transport. However, similar to many other commodities, its supply takes years, or even decades, to shift. Brazil have and has had for more than half a century an electric system based on hydroelectric power. This brings the tremendous advantage of environmental sustainability, but also makes the forecasting of supply quite erratic and dependent on variables as uncontrollable as the weather.

In recent years, a vast portion of the Brazilian territory has seen Its water flows being drastically reduced by a drought that had no equal since the early 1950's. Such portion includes to a large extend the southeast, where the largest cities – and thus the largest consumption centers – like São Paulo and Rio de Janeiro are located. As supply fell short, a very substantial increase in electricity price took place – including spot prices surges from levels of BRL 30 to 60 to over BRL 600. In January 2015, the government slashed the spot prices, creating an artificial ceiling of BRL 388.04. Since then, the prices remained at the ceiling, and investors in generation are extremely optimistic that the drought will keep the prices at such high level for a period of 3-5 years after January 2015 (Small Hydropower Investors, 2015).

With such optimism, investors are focusing their capital into expanding installed capacity of electricity generation. Those residing in regions less affected by the drought are still focused on Small Hydropower Plants, or PCH's – from the Portuguese "Pequena Central Hidrelétrica". As witnessed on the interviews performed for this thesis, investors that have Small Hydropower Plants going into operation right now are not infrequently leaving the majority, or even the totality, of their new plants' energy to be sold on the spot market. This behavior surely has been profitable over the last year, when spot prices have been about twice as high as the price on the longer-term free market; however, should the prices decrease from the ceiling, it is expected that longer-term prices should also be reduced, thus leaving investors exposed within the lifetime of their assets – i.e. a minimum of 30 years since the beginning of the plant operation. With prices being re-evaluated and redefined by the Chamber of Commerce of Electrical Energy (CCEE) on a weekly basis, this position of dedicating most of the portfolio to selling on the spot market points to a very short-term, potentially risky investment behavior.

With an opposite stance, 48 out of the 472 (ANEEL, 2015) (CCEE, 2015) existing PCH's in Brazil have entered auctions to sell energy on the regulated market, with 45 of them selling their future production, for 30 years, prior to their construction. Furthermore, 32 of those sold more than 75% of their total production, meaning they deliberately chose to lock their assets into a position of selling their production of 30 years under a price specified at the moment of the auction, only to be readjusted by inflation.

As an important note, the goal of this thesis is not to forecast prices or the behaviors of agents of the Electrical Energy. Neither it is to point towards a single solution of optimal portfolio of electricity sales. Not only such exercise would unlikely create value beyond of mere speculation, but also the past behavior of investors towards full commitment into either the shortest term or the longest term options are indications of potential room for improvement in the structuring of their portfolios. Therefore, this thesis aims to provide insights in terms of the impact in the expected present value of revenues that different portfolio choices could have within different future scenarios, as well as their respective standard deviations. Furthermore, it also aims to explore the current situation and trends of the market, including the impact of the drought and the planned increase in supply by the Brazilian government.

5. Part II – Hydroelectric Generation

5.1. UHE vs. PCH – Large versus Small Hydropower

5.1.1. Capacity and Size of investment

While PCH's are limited by law to the size of 30MW in installed capacity (Brazilian Government, 1998), the larger Hydroelectric Power Plants, also known as UHEs, can be as large as several thousand Megawatts. Itaipu, the binational hydroelectric built in joint venture by Brazil and Paraguay, is the largest in the region and second largest in the planet, with 14,000 MW. The average size of a PCH in Brazil is 12 MW, considering the 448 plants in ANEEL's database. For UHEs, there is a lot more variation in installed capacity, with the average being 598 MW per plant within the 149 UHE plants with more than 30MW. The distribution of UHEs in terms of installed capacity is better observed in Graph 31.

More interesting though, is the insights derived from looking at the factor of Physical Warranty divided by the installed capacity, which we are referring to as Utilization Rate. As UHEs tend to have large reservoirs, and as PCHs are often "at river-flow" (i.e. with no reservoir), it would be expected that they would have a larger utilization rate. However, analyzing ANEEL's plants' data, the 448 PCHs have an average of 58% with a 15% standard deviation, while the 148 UHEs over 30MW have an average of 55% with a standard deviation of 16% (ANEEL, 2015). However, what is really important from the investors' perspective is to understand the investment size per MW of Physical Warranty. This data on utilization rates is better observed in Graph 32, while the distribution of installed capacity in PCHs can be seen in Graph 33.

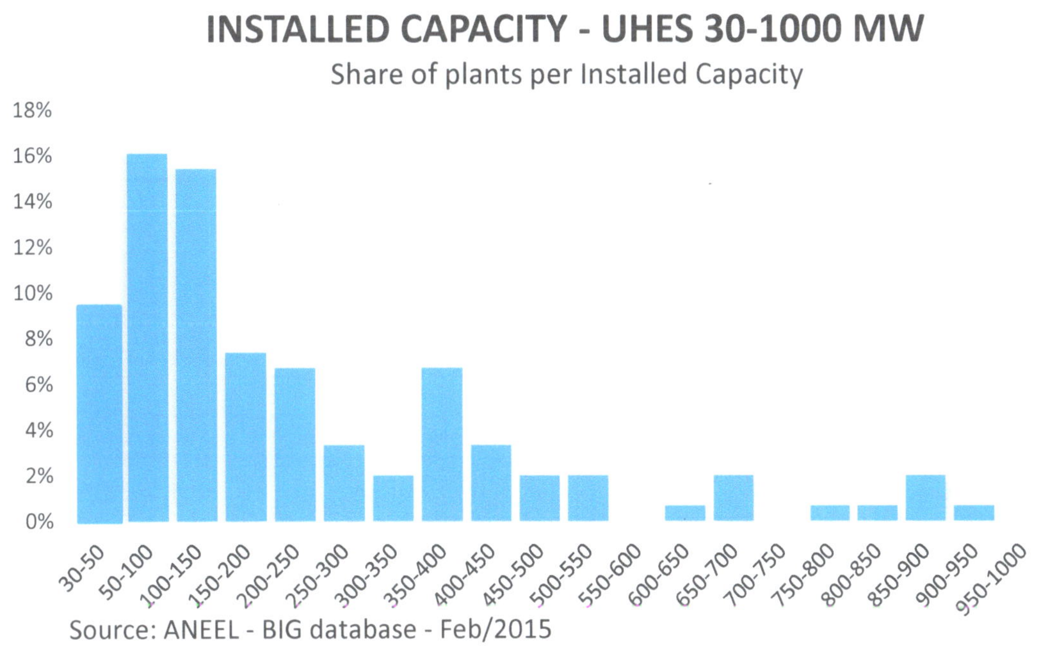

INSTALLED CAPACITY - UHES 30-1000 MW
Share of plants per Installed Capacity

Source: ANEEL - BIG database - Feb/2015

Graph 31 - Installed Capacity of UHEs

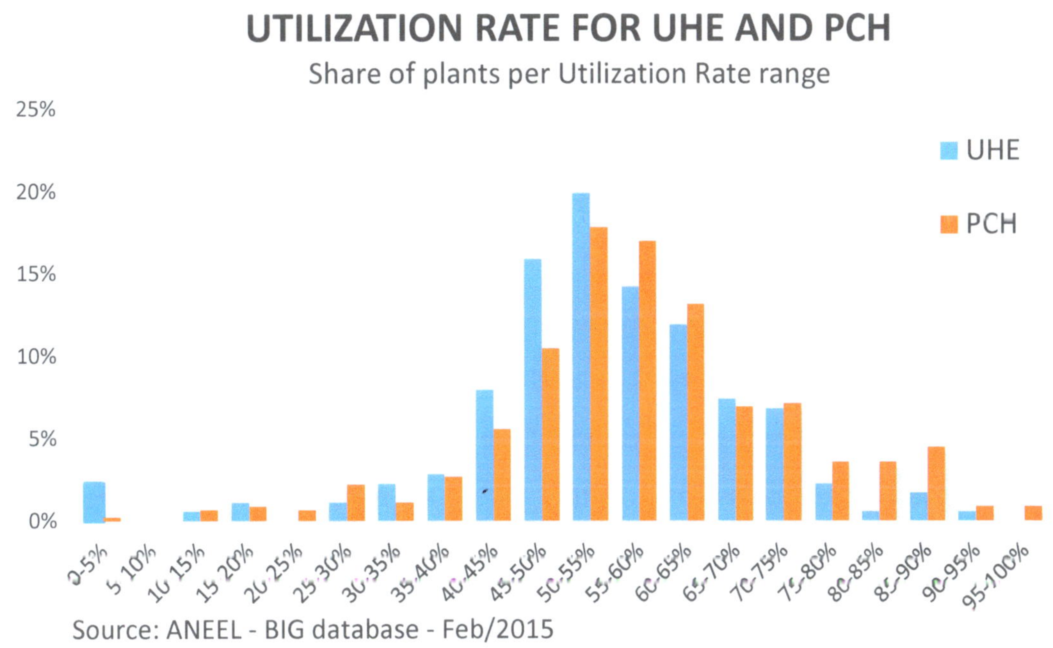

UTILIZATION RATE FOR UHE AND PCH
Share of plants per Utilization Rate range

Source: ANEEL - BIG database - Feb/2015

Graph 32 - Utilization Rates for UHEs and PCHs

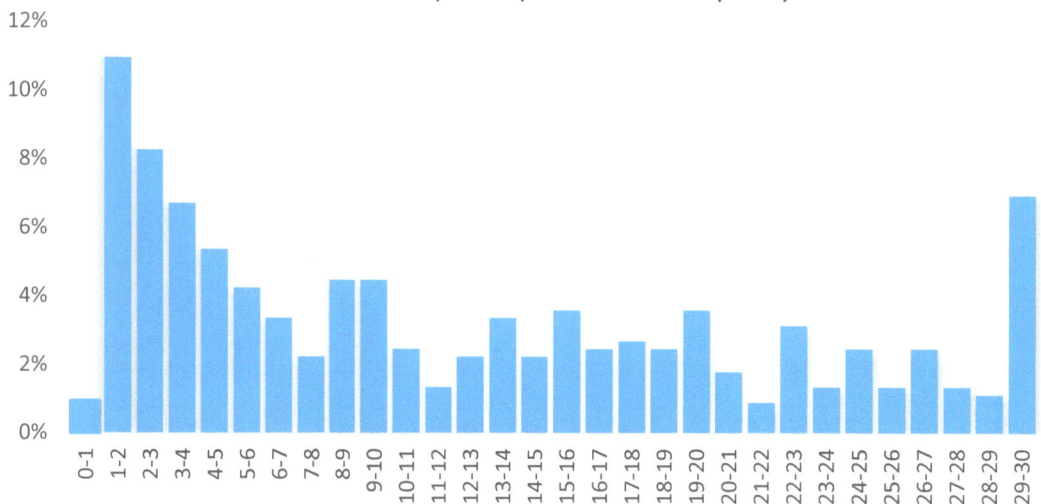

Source: ANEEL - BIG database - Feb/2015

Graph 33 - Installed Capacity of PCHs

The data for investment of each plant is available at the "Resultado Consolidado de Leilões", a monthly-updated document by CCEE that has extensive data on plants that enter auctions. However, out of the 48 PCHs and 99 UHEs listed, only 26 PCHs and 10 UHEs have investment data (CCEE, 2004-2015). Within this limited dataset, the average investment per MWa (i.e. MW of Physical Warranty) is BRL 11.0 Million for PCHs and BRL 7.4 Million for UHEs, as shown in Table 2. As expected, there is an average ~30% gain in efficiency of the investment when moving from PCHs to the high-scale UHEs. On a very important note, this is considering the expected investment, at the date of the auction – PCHs are much less likely to suffer delays due to their minimal environmental impact and to their low visibility on the media. This data is shown in Graph 34. In other words, investors need to take into account that this 30% of efficiency gain is at risk if the plant does not enter the market within the expected time. Such delay may expose the plant to the PLD, since it would have to purchase energy to supply an equivalent amount of the production it was supposed to have during the delay period. As this happens in the beginning of supply, and as interest rates are high in Brazil, this may have a significant impact in the NPV and needs to be taken into account by UHE investors. Lastly, the much higher complexity of building a very large UHE also brings more uncertainty in terms of the investment required. For instance, the UHE Belo Monte was expected to cost about BRL 19 Billion at the date of the auction, but it is already expected to cost over BRL 30 Billion, over 57% more than forecasted (VEJA Magazine, 2013). Although unnecessary to detail here, increases in expected investment are not unusual and can also be seen in other large-scale UHEs, such as Teles Pires, Colíder, Santo Antônio or Jirau.

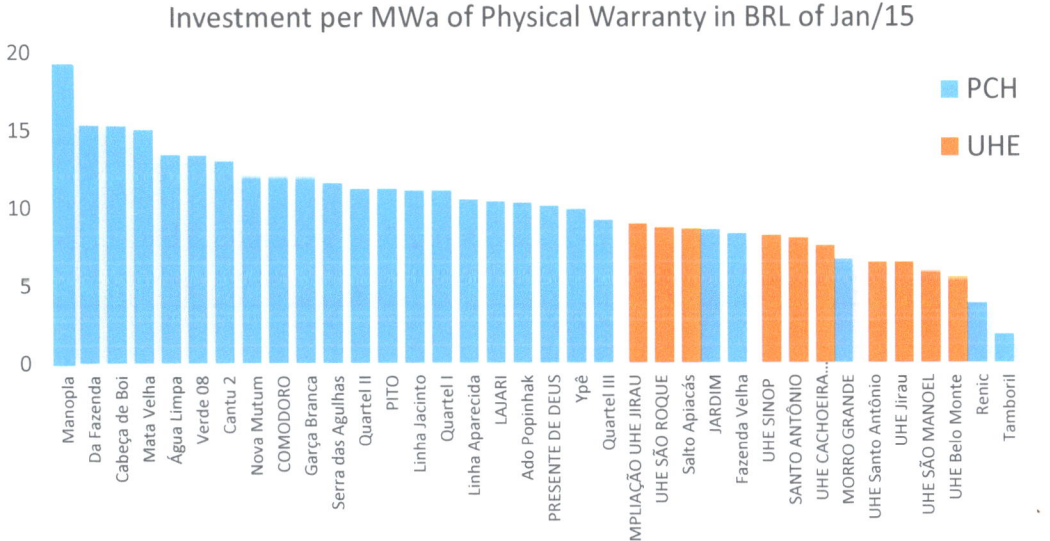

Graph 34 - Investment per Average MW of Physical Warranty

Table 2 - Investment in PCH and UHEs, from CCEE, in BRL of January/2015

		PCH (26 plants)	UHE (10 plants)
BRL/MW	Average	5.691.812	4.000.226
	Std. dev.	1.951.899	773.776
	Min	802.782	2.246.633
	Max	9.919.140	4.901.841
BRL/MWa	Average	10.995.449	7.414.758
	Std. dev	3.525.345	1.249.094
	Min	1.823.825	5.521.035
	Max	19.075.269	8.910.937

5.1.2. The ACR versus ACL Flexibility for Portfolio Optimization

One of the most important benefits of PCHs versus large UHEs is the possibility to sell energy in a very flexible portfolio. A PCH investor may decide to sell the totality of its energy on the ACR, for 30 years of operation, prior to construction. The same investor may also decide to sell 100% of its generation via bilateral agreements, undisclosed to the public, with pricing and period terms decided at will in common agreement between the two parts. Furthermore, there is also the possibility of leaving energy to the PLD clearing prices, as well as to join the MRE protection pool or not. Lastly, the investor is free to pick a combination of such choices, whether joining or not the MRE. This strong flexibility allows the PCH investors to optimize the portfolio in the way they decide to be best, as well as to adapt over time (as long as they did not commit energy for a very long period prior to construction).

On top of such flexibility, PCH investors are not required to compete bidding on the lowest price to be sold at the ACR in order to gain the rights to construct the plant. Not only the sales at the ACR are facultative, but also whenever they are done, PCH investors depend only on finding an ACR buyer (e.g. a distribution company) willing to pay the price they offer. For instance, it has been recorded that the PCHs of Tombos, Fagundes and Areal have sold energy at the ACR in 2014 for BRL 270/MWh, while in December of 2013 the consortium by EDP and Furnas (the latter which is part of Eletrobrás) won the right to build the UHE São Manoel by committing to sell energy at BRL 83.49/MWh.

To win in the UHE São Manoel competition, said consortium bid a price with 22% discount over the ceiling-price of BRL 107/MWh, which shows the impact that several consortiums competing for the same power plant may have – in the case of the UHE São Manoel, there were 5 of them. To be unbiased in terms of dates and the fluctuation of price over time, within the same auction of the UHE São Manoel, the average price sold was of BRL 109.93, ~24% higher than the EDP/Furnas consortium bid. Moreover, the average price for PCHs within this same auction was BRL 137.35, a price ~64% higher than the one of said bid (Exame Magazine, 2013). As a consequence, since 2007 all winning bids of UHEs have been selling the minimum ACR energy established by the auction rules, as it could be seen on UHEs such as Santo Antonio, Jirau, Baixo Iguacu, Belo Monte, Garibaldi, Teles Pires and others.

As opposed to the PCHs flexibility, the large hydropower plants are also often restricted by the rules of the auction where they are released. Since they have structural importance for the country, the government has to ensure that they serve the purpose of supplying to all final usages, which means to a large extent the necessity of participating at the ACR. Prior to each auction, ANEEL releases a specific "Edital", a public notice that details the general and specific rules for that auction. In such document, whenever a new UHE construction participates, there are rules that limit the sale of energy by the bidders. A common setting for this, for instance, is to demand bidders to sell a maximum of 10% of their Physical Warranty on the ACL, a maximum of 20% to self-production (i.e. if among the bidding consortium there is an equivalent equity ownership by a large consumer that will purchase the energy), and a minimum of 70% to the ACR. As described before, the bidding for UHEs is the based on the price that the plant will sell at the ACR, meaning that a minimum of 70% of the produced energy will be sold for 30 years at the lowest price bid (i.e. the winner of the auction).

5.1.3. Players Involved

The nature of UHEs and PCHs in terms of size also determines the kinds of players that are able to compete within each one of the segments. For PCHs, a group of investors with BRL 5-10 Million may already consider enter the market with a small plant. However, as reported in one of the interviews for this thesis, one of the barriers to entry in this market is the "learning curve" of PCH construction. In this interview, the investor said that their current and fourth plant was costing about half per MW of capacity than their first one, although the conditions were very similar in kind. Considering this, a new investors should be willing to invest more than in just a single plant in order to make it viable for them to enter the market. Due to these reasons, it is possible to see a wide variety of companies owning PCHs and thus a highly fragmented market, even though large players like CPFL or CEMIG also have investment in such plants.

As for UHEs, the set of participating companies is much more limited. As said before, Eletrobrás and its subsidiaries are often part of the winning consortium, usually together with at least one energy-focused company, like EDP, CPFL, CEMIG, AES, Copel, NeoEnergia, Suez and Tractebel. With a more significant presence on the larger UHEs, construction companies are often part of such consortiums as well, as the investment required deeply affects the Net Present Value of the UHEs and is largely dependent on the performance of said companies. The construction companies seem in UHEs are mostly Odebrecht, Camargo Corrêa, Andrade Gutierrez, Queiroz Galvão and Engevix. Due to recent incentives towards self-production, large consumers like Vale (mining) and Alupar (Aluminum) are often part of such consortiums as well.

5.1.4. Other Considerations

On top of the advantages mentioned above, there are also other considerations worth mentioned about Small Hydropower Plants. The first one is regarding the environmental impact of their construction: with its "fuel" being the natural flow of water, and without the large reservoirs required by UHEs, PCHs are one of the most environmentally-friendly ways of generating electricity. Figure 6 was published by McKinsey & Co., and it shows the performance, in terms of tons of CO_2 increased/reduced per Euro invested, of many types of investment in terms of energy generation/efficiency. Small Hydropower has a negative position, meaning its performance in such metric is so good that each Euro invested implies in decreasing the CO_2 in the atmosphere. Furthermore, PCHs are also ahead of geothermal energy, Solar Photovoltaic generation and Wind power, meaning it is one of the best electricity generation forms known according to this metric.

Exhibit 6

V2.1 Global GHG abatement cost curve beyond BAU – 2030

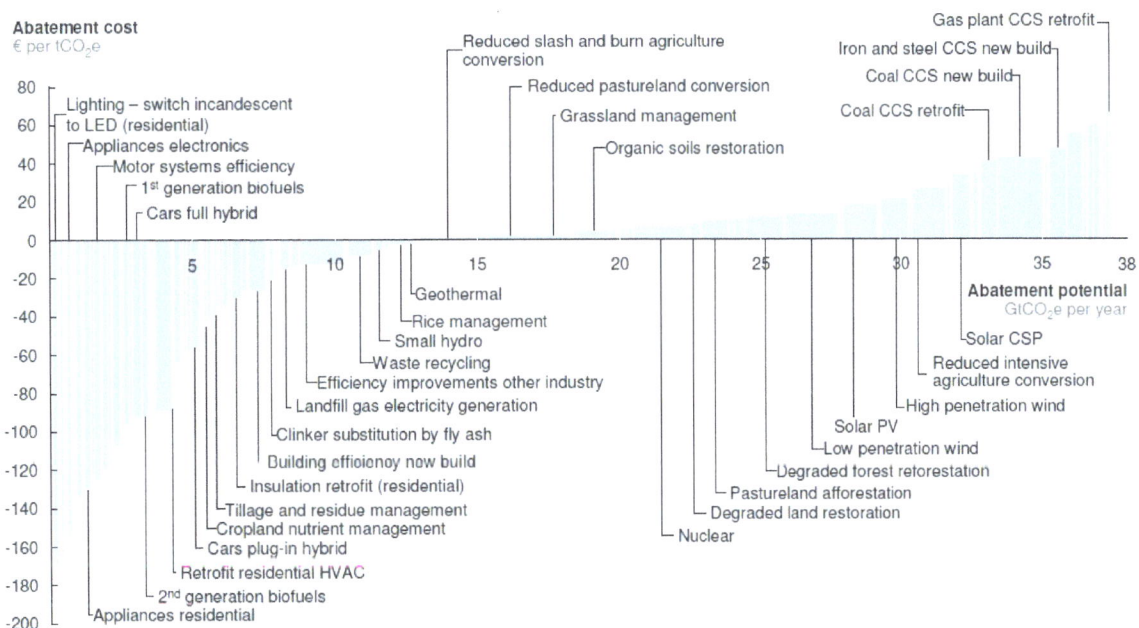

Figure 6 - McKinsey chart on Global GHG abatement cost curve (McKinsey and Company, 2014)

Furthermore, the Brazilian market is known for having very difficult environmental licensing frameworks, often causing significant losses and delays for investors in large capital expenditure projects such as power plants, mining and others. As PCHs are small and have little effects on its surrounds, this process is, although far from easy, much less complex than for UHEs. In section 4.2 there is a brief explanation on the process required to receive the LP, LI and LO, the respective licenses needed for project, installation and operation of the PCH.

One last interesting consideration on PCHs is the timeline of their project and construction. As investors are not required to win the rights for a PCH through auctions, as is the case with UHEs, the timeline for project management, decisions and construction is much more flexible. An investor researches opportunities to build PCHs, decides when to start and how to sell the production at his or her own pace. In times of capital constraints, investors may simply defer a project – as opposed to a UHE, which due to its structural importance for the country has to participate in the auction according to the government's will.

5.2.   The Overall Process of Investing in a PCH

The process of investing in a PCH is far from trivial – yet, it is well known and formalized, creating efficiency gains opportunities for recurrent investors. The process starts by searching for a river inside a private land that has a propitious site for a Small Hydropower installation. With a rough first estimate of the potential of the hydro, the PCH investor has to negotiate with the landowner in order to obtain authorization to install the plant. Once they agreed on a deal structure, which could go from actual purchase of the land to profit-sharing of the PCH operations, the investors start the project phase. The following process has an overall flowchart as seen in Figure 7, extracted from the PortalPCH.com.br website (PortalPCH, 2014).

The first step is then to analyze in detail the potential of the plant. If such potential is not fully understood, the investors have to go through the "Simplified Inventory" process, as described in the resolution number 393, released by ANEEL in December of 1998 (ANEEL, 1998). This resolution established the general procedures for registration and approval of hydroelectric inventory studies of watersheds. Following this process, there is a first evaluation of the viability of the plant. Should this evaluation show an interesting potential from the perspective of the investors, they move forward with the project; otherwise, the project is archived.

If this first evaluation shows an interesting potential of hydropower generation, the next step to further detail the project by (1) acquiring in-site data through measurement and research, (2) performing the basic studies of implementation, (3) creating a preliminary lay-out of the plant, and (4) estimating the budget required for the project. With this further detailing, the economic viability of the project is tested. Again, should it be interesting enough, investors move forward; otherwise, they archive the project.

With a positive conclusion from the economic viability study, two simultaneous processes have to be undertaken. The first one is to register the project to execution at ANEEL, which is followed by a detailed elaboration of the engineering project, hydrological and geological measurements, as well as energy and transmission studies. The second process refers to the environmental impact of the installation, and includes both the elaboration of the EIA/RIMA documents ("Study of Environmental Impact" and "Report for Environmental Impact"). The final objective of this second process is to obtain the "LP", or "Preview License", as it is called the first of the three key environmental licenses of the project.

With a LP in hands, the investors move forward to optimize the engineering project of the plant. Then, three simultaneous process take place, with a single final object of preparing the construction of the plant. The first one is to present the detailed project to ANEEL, together with the LP, in order to get it approved. If ANEEL decides not to approve the project, it will determine requirements that have to be met. Once the investors change the project and meet such requirements, they apply again for approval.

The second process in this phase is to elaborate the "PBA", which is the environmental project that will be executed during the installation of the plant. The PBA is used by the investors to apply for the LI, the "Installation License" and second of the three key licenses. In the same way ANEEL does, the environment regulators either approve the project and issue the LI, or deny it and make a list of requirements that have to be met. Once investors change the PBA to include the requirements, they apply again, repeating this until they get the LI.

The third simultaneous process is to apply for the "Outorga de Uso da Água", a governmental grant that allows investors to utilize the water to generate power. In the same way as the other two process, either the government issues the grant or it denies it while making requirements. Once requirements are met, the government issues the grant.

Finally, with the project approved by ANEEL, and with the LI and the Water Usage Grant issued, the investors execute the project, building the plant while complying with all requirements. Once the plant is built, if all requirements were met, the investors obtain the final license - the LO, which is the "Operating License" required to run the PCH and provide power to the market.

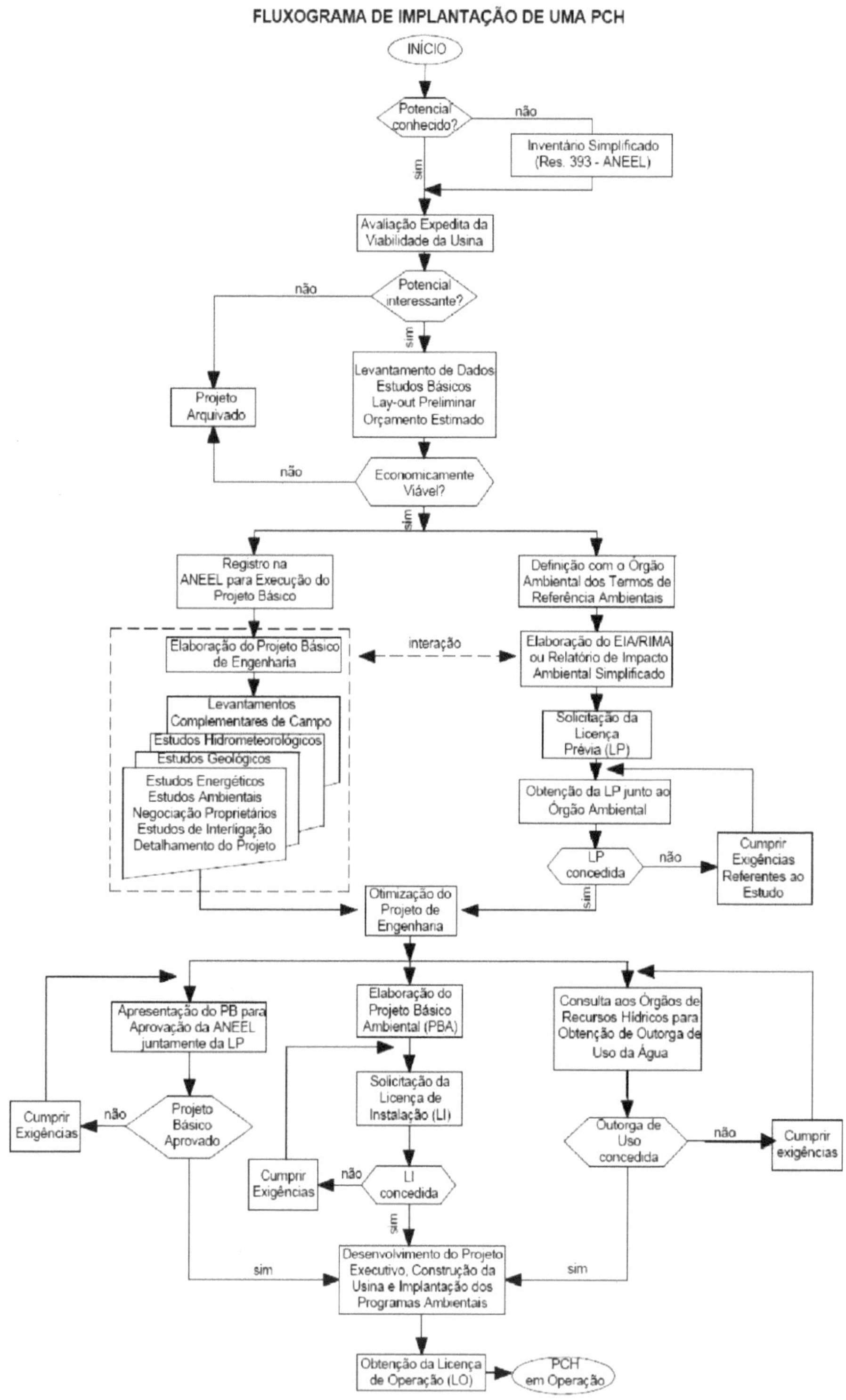

FLUXOGRAMA DE IMPLANTAÇÃO DE UMA PCH

Figure 7 - Flowchart from website "http://www.portalpch.com.br/" from project to operation

5.3.   Installed Capacity and the Physical Warranty of a PCH

The Installed Capacity of a PCH, also known as "Nominal Capacity", refers to the installation of the plant. A very large plant may have a high installed capacity, but if there is insufficient water flow, it will not be able to utilize its installation to the full power. Due to the high variability of hydropower installations, from size of reservoirs to regional seasonality of water flows, the ANEEL utilizes a measure called "Physical Warranty" to determine the average expected generation of a plant. The installed capacity is rarely reviewed, but the Physical Warranty is constantly observed by ANEEL. In a year of drought, Hydropower plants will produce significant less, and thus will have their Physical Warranty reduced by ANEEL. A "Utilization Rate" is thus a measure of how much of the installed capacity is actually being utilized to generate energy, and is given by the division of the Physical Warranty by the Installed Capacity. This measure can be observed for a wide range of Brazilian Small Hydropower Plants in Graph 35.

Often, investors measure their investments' construction performance in terms of Millions of BRL per MW of installed capacity. In most cases, a healthy utilization rate would be between 70-85%, but as ANEEL data shows the average of the country is far below that, at 58%, with a high standard deviation of 15 percentage points. Causes of this variation might be the average age of the plants, the current drought (as the Physical Warranty data is from February/2015), project mistakes and so on.

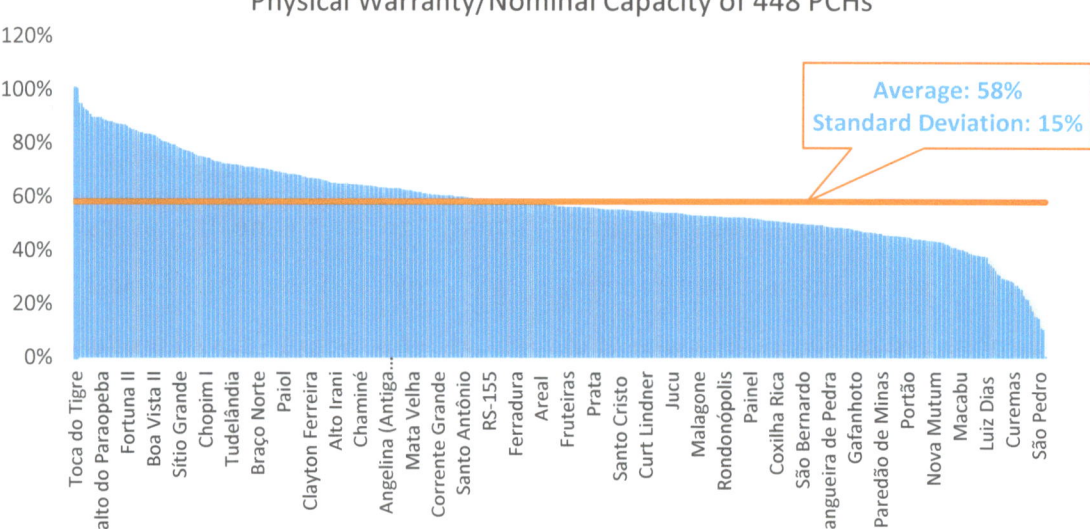

UTILIZATION RATE OF PCHS
Physical Warranty/Nominal Capacity of 448 PCHs

Average: 58%
Standard Deviation: 15%

Source: ANEEL - BIG database - Feb/2015

Graph 35 - Utilization Rate of PCHs

It is also important to notice that the Physical Warranty in question is an annual measure. This means that a plant with 10MWa of Physical Warranty is expected to produce 10 MWa * 8,760 hours = 87,600 MWh in this single year. However, many PCHs don't have reservoirs, and are strongly affected by the month-to-month seasonality throughout the year. This means that in the rainy months of February and March they may produce significantly more than in the dry months of August and September.

Utilizing data from CCEE of 397 plants, the average production of a PCH per month is as shown in Graph 36. This data is useful in two ways: first, it is useful to understand the average impact of monthly seasonality on a Brazilian PCH; second, it will be useful as input to the monthly model created at Part III of this thesis.

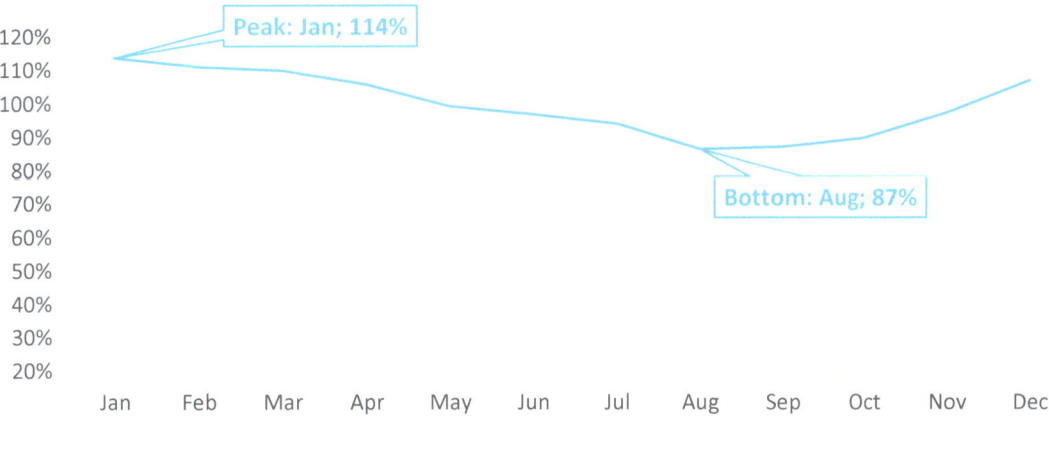

MONTHLY PHYSICAL WARRANTY OF PCHS

Average monthly production in share of Physical Warranty, based on
397 PCHs - in Monthly MWa/Physical Warranty in MWa

Peak: Jan; 114%

Bottom: Aug; 87%

Source: CCEE

Graph 36 - Average Expected Monthly Physical Warranty of PCHs

Lastly, the same dataset also shows that deviations from the expected generation occur with a normal behavior. Graph 37 ahead exemplifies this behavior through the utilization of 32,295 data points on monthly deviation of the 397 plants. Furthermore, the data on Graph 37 was gathered from all months of the year. On the model described in Part III of the thesis, the standard deviation considered excludes outliers and was acquired for each month of the year, in order to better model the month-to-month seasonality of PCHs. This is described in further detail in section "6.2. Modelling the Portfolio".

MONTHLY GENERATION AS % OF PHYSICAL WARRANTY OF PCHS

Normalized behavior of plants' actual generation in relation to the expected generation (Physical Warranty = 100)

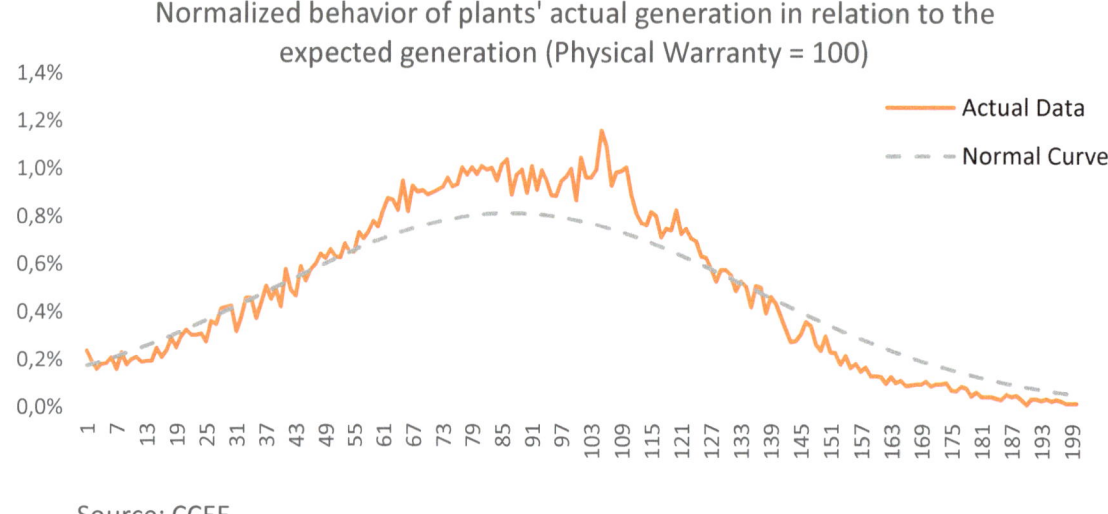

Source: CCEE

Graph 37 - Monthly Generation as % of Physical Warranty of PCHs

5.4. The MRE: an Energy Relocation Mechanism to Reduce Risk

As described before, the MRE works like a pool that allows PCHs to reduce the financial risk inherited from the hydrological variation impact on the electricity generation. With the current drought, 2014 became a great year to illustrate how powerful the MRE can be for PCHs. Graph 38 shows a comparison of the expected monthly rate of production (MWa) as a percentage of the annual Physical Warranty, as in Graph 36 of section "5.3. Installed Capacity and the Physical Warranty of a PCH", with the same measurement of the actual production in 2014.

The production of the 339 plants with complete datasets for 2014 at CCEE's sources was 81% of what it was expected to be (CCEE, 2015). As for the country, this impact is not incredibly relevant, as this means 4TWh less in the market – or ~0.7% of the total consumption. However, for some PCH investors, this could mean bankruptcy.

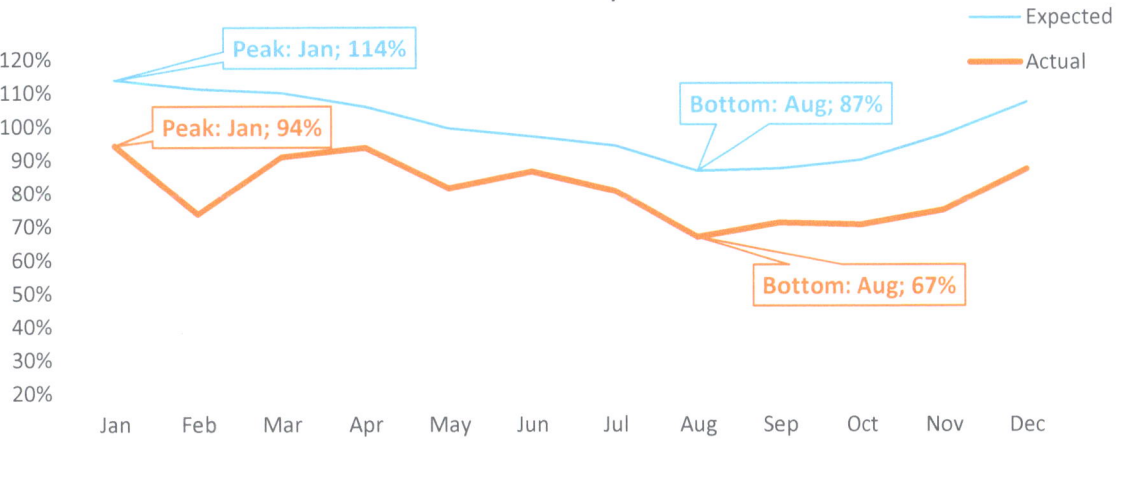

Graph 38 - Expected Monthly Physical Warranty of PCHs versus 2014 Actual Data

With a Physical Warranty of 2.4GWa, those 339 plants together were supposed to produce 21.2 TWh of electricity. In 2014, they produced 17.2 TWh, or 81% of their expectation. To estimate the financial risk that this would represent with and without the MRE, it is first necessary to understand how much less electricity than expected was generated each month. Then, to research the monthly values of the TEO (i.e. the price paid by generators to "complete" the energy they sold on the MRE) and the PLD (i.e. the same, but for plants out of the MRE).

As Table 3 shows, the TEO of 2014, determined in November of 2013 (ANEEL, 2013), was BRL 10.54 (nominal values) for every month of the year. The PLD, on the other hand, varied from BRL 387 to BRL 840 (Jan/2015 real values). This means that a PCH within the MRE paid BRL 10.54 (or between 10.7 and 11.3 in BRL of Jan/2015) to purchase each MWh that it lacked in generation but had already sold – and this price was paid to another plant within the MRE that had the fortune of producing more than its Physical Warranty. A PCH that was outside the MRE, however, paid anywhere between BRL 387 and BRL 840 to a very fortunate electricity producer that did not sell the entirety of its production beforehand. As it can be observed in the calculation below, if the whole pool of 339 plants measured was outside the MRE and exposed to the PLD, they would have paid BRL ~2.8 Billion to acquired electricity to honor their sales. Should they all have been inside the MRE, this cost would have been BRL 44 Million, or about 1.5% of the latter.

Lastly, it is highly recommended that any investor or student of the Small Hydropower generation market in Brazil should deeply understand the details behind the MRE. A great starting point and complete document can be found at CCEE's website, called "Mecanismo de Realocação de Energia", with the latest version being from 2013 (CCEE, 2013).

Table 3 - Monthly generation and simulation of TEO versus PLD exposure for 339 plants of CCEE's database of 2014

Month (2014)	Expected Production (TWh)	Actual Production (TWh)	Generation Lack (TWh)	TEO (BRL Jan/15)	PLD (BRL Jan/15)	Losses within MRE (BRL Jan/15 Millions)	Losses out of MRE (BRL Jan/15 Millions)
Jan	2,1	1,7	0,4	11,29	386,97	4,0	137,0
Feb	1,8	1,2	0,6	11,21	773,04	6,9	472,6
Mar	2,0	1,6	0,3	11,11	815,21	3,8	281,8
Apr	1,9	1,6	0,2	11,04	828,74	2,4	177,8
May	1,8	1,5	0,3	10,99	715,24	3,6	231,1
Jun	1,7	1,5	0,2	10,94	398,23	2,0	73,6
Jul	1,7	1,5	0,2	10,94	560,12	2,7	138,0
Aug	1,6	1,2	0,4	10,92	712,46	3,9	255,6
Sep	1,5	1,2	0,3	10,85	759,78	3,1	215,7
Oct	1,6	1,3	0,4	10,81	776,08	3,8	273,1
Nov	1,7	1,3	0,4	10,75	839,53	4,3	332,2
Dec	1,9	1,6	0,4	10,67	643,52	3,9	232,6
TOTAL	21,2	17,2	4,0			44,2	2.821,2

5.5.   The Role of UHEs and PCHs in Brazil

From a single company's perspective, investing in Small Hydropower can be a powerful tool to play on the free market. Such plants allow a fairly stable supply, flexible sales options, simple and low-cost operations, strong protection against spot price surges through the MRE and the ability to build a single plant with anywhere from BRL 5 to 150 Million. This also means that an investor can execute several similar constructions with smaller investments, gaining capital expenditure efficiency quickly while utilizing resources. On the other hand, large electricity players such as distribution companies might not see enough scalability on Small Hydropower

Plants, and may then opt to build larger hydropower facilities. UHEs allow investors to execute a much lower number of projects, while obtaining a usually favorable investment/MW ratio. Furthermore, UHEs also ensure lower prices at the regulated market, thus it is of interest of distribution companies to make sure such auctions remain competitive. It also corroborates this last point that most large hydropower consortiums have strong participation of such distribution companies.

From the perspective of Brazil as a country, Small Hydropower brings great benefits, such as minimal environmental impacts, lower barriers of entry for the electricity supply market due to the smaller investment size, and the ability to increase supply in fewer years than through larger hydropower plants. However, as demonstrated in the first part of this thesis, even in very high numbers Small Hydropower Plants represent a small fraction of the electricity supply needed. As an example, it would take approximately 470 of the largest PCHs to supply the same power of a single Itaipu UHE. In fact, according to ANEEL's annual report, Small Hydropower represented 4.8 GW of the 133.9 GW of the installed capacity by December 31st, 2014 (ANEEL, 2015). In the same date, UHEs represented 84.1 GW of capacity. As references, the current largest hydropower plant, Itaipu, has an installed capacity of 14 GW, while the largest one being built, Belo Monte, will have a capacity of 11.2 GW. Thus, from the country's point of view, PCHs are interesting and efficient, but UHEs are the ones that ensure that the large part of the demand is being met.

Anyhow, even while not having enough scale to cover all the electricity demand growth in Brazil, Small Hydropower is of vital importance for the country. The government has been well aware of that for a long time, which is demonstrated by its promotion of such investments through major subsidize programs such as PROINFA. The Brazilian government established the "Programa de Incentivo às Fontes Alternativas de Energia Elétrica", or "Program of Incentive for Alternative Sources of Electrical Energy" in 2004, through the decree number 5,025 (MME, 2004). Its objective was to increase the share of Wind power, Biomass and Small Hydropower in the total electricity production in Brazil. To do so, it added an extra layer of tariffs, commonly referred to as "Proinfa", for most of the electricity consumers in Brazil – the only exceptions being low-income residences and self-producers – aiming at gathering resources for the subsidies. This program successfully subsidized the construction of 144 plants, in a total 3.3 GW of installed capacity. Of this capacity, 63 Small Hydropower Plants represented 1.2 GW, 54 Wind Power parks represented 1.4 GW, and 24 biomass facilities represented 0.7 GW. All the energy generated by these facilities have a 20-year guarantee of purchase by Eletrobrás, the mixed-economy enterprise that owns significant assets in generation, transmission and distribution in Brazil, as described in section "4.3.6. Eletrobrás".

5.6.   A Topic for Further Study: Wind and Hydro Complementarity

During the interviews, some investors mentioned that wind and hydro-power generations have complementary monthly generation, meaning that an investor with a good portfolio between the two sources would have a more "stable" monthly generation throughout the year. Although this is outside the scope of this thesis, this is an interesting theory. Yet, even with some indications in its favor, it is just a theory that might be true in general, within a region, within a small group of specific plants, or even not true at all.

Economic-wise, this could be interesting in terms of reducing monthly cash flow variation – i.e. months would have more similar revenues when compared to each other. However, even if it would be true, it would not impact the standard deviation of the expected revenue of a single month, unless the generations would be complementary in the sense that if one lacks energy, the other produces in excess. This is a different complementarity, that so far there is no indication of existing. In other words, the complementarity investors observed shows that the expected revenue (the "mean" of the revenue) would vary less month-to-month. The more interesting impact, however, would be from showing that when hydro lacks energy, wind produces in excess, and vice-versa. In the latter case, even if the expected revenue would vary more, its standard deviation would be reduced – investors would have a much better ability to forecast their revenues.

To illustrate the kind of "complementarity" that has been observed, one could read the report of the company XPGestão, from Q3 of 2014 (XPGestão, 2014). This reports talks about Asteri Energia SA, a holding that owns 100% of the wind-power facility Gargaú, and 51% of the

78

PCH Pipoca. The report shows that the month-to-month generation of both plants combined is more stable than any of them alone, but more interestingly it shows that while the wind-power produced 15% more than expected in 2014Q3, the PCH produced about 50% less than expected in the same period. How this latter observation would remain true over time, with statistic relevance, and if that is a general attribute of wind and hydro power facilities located under the same weather, could be topic for very interesting studies. This thesis does not explore this topic in any further detail.

6. Part III – Electricity Sales: A Portfolio decision

6.1. The Most Important Decision of a PCH: When and How to Sell Electricity

The profile of the investment in a Small Hydropower Plant is interesting: a large initial capital expenditure, leading to a simple operation of an asset that lasts a minimum of 30 years and provides a commodity to the market with very low variable costs. Therefore, optimizing operation of the plant itself is not nearly as important as the two key drivers of value for investors: the execution and optimization of the capital expenditure and the optimization of the sales of electricity. The former will not be further detailed in this thesis, although it is indeed highly important to increase the expected net present value of the investment. The latter depends on the decision of how the plant will sell electricity, thus on the creation of a sales portfolio.

As described earlier on this thesis, there are many ways of selling electricity in the Brazilian market, from the clearing-price of the PLD to the 30-year contracts of the ACR. Each of these depend on decisions made by the PCH investors, and such decisions have to be made according to a certain timeframe in the lifetime of the plant, as described in section 5.2. Such methods vary significantly in (1) the length of the contract, thus the length of the period where the contracts guarantee the price per MWh; (2) the variation (i.e. average and standard deviation) of the price being practiced on the market; (3) periods in the lifetime of the plant where the investors are allowed to sell through the given method. While the optimization of the capex can be well planned and depends significantly on "learning-by-doing", the sales portfolio is a value driver that depends on investors' risk appetite and on the market. Although "depends on the market" is a very general statement, the PCH sales long-term performance depends externally more specifically on (1) how fast the demand for electricity grows; (2) how the long-term increase in the country's electricity supply is planned and executed; (3) how the weather supplies water to the large hydropower plants in Brazil, which ultimately determine the shorter-term marginal cost of supply and thus heavily influence prices practiced on the market.

Lastly, as capital expenditure execution is a matter of optimization of operations, creating the right portfolio of sales is ultimately the most important decision for a Small Hydropower investor in Brazil. With this in mind, this thesis follows next by creating and describing a model to simulate the difference in the present value of revenues given a few possible sales portfolio combinations, more specifically to illustrate the extent of the impact of such decisions (section "5.2. Modelling the Portfolio"). Later on, it continues by detailing the alternatives within the portfolio, including some observations on their past behavior (section "6.3. Possible Revenue Sources: The Decision Flow and its Timeline"). Lastly, sections 5.4. to 5.7. will discuss insights derived from the results of the simulations of the model.

6.2.   Modelling the Portfolio

The model was built considering three types of variation within a single base-model. The base structure as well as each of the variations are described within this section. The first variation is the standard deviation of the actual generation in comparison to the Physical Warranty. Datasets from 397 plants were analyzed (CCEE, 2015), and since the level of variation can be significantly different for PCHs regardless of size, this was a variable considered on the simulation. The second variation considers that investors that entered in the recent past or entering in the short future are joining during a time of drought, which maintain PLD prices at their ceiling. This variation considers the length of the continuation of the drought from the beginning of the operations. Lastly, the third variation considers 8 different types of portfolios, in order to compare their performance in terms of expected present value and standard deviation of future revenues.

The results of the crossing of these three types of variation led to the construct of 96 different simulations ran through 1,000 iterations each, from which the results in the following sections were derived. The final goal of the simulations and the analyses of their results is to have a better understanding of the impact of the variables described in this section on the present value of future revenues. The software utilized was Microsoft Excel, with Palisade's add-in @Risk.

6.2.1. The Structure of the Model

The Structure of the model is divided in three steps, as shown in Figure 8. The first step is to simulate the actual generation of the plant, based on the expected generation and its standard deviation. The expected generation considered a plant with 10MW of installed capacity and 80% utilization rate, thus with an 8MWa Physical Warranty. This means that the plant is expected to produce 70,080 MWh in each year. This production had its seasonality considered month-to-month, with expected values aligned to the findings described in Graph 36 of section "5.3. Installed Capacity and the Physical Warranty of a PCH". The stochastic behavior of the actual generation obeys a normal distribution, which is realistic according to the past performance data on 397 plants (CCEE, 2015).

The second step of the model is to transform the generation into monthly revenues throughout the 360-month period covered by the model. To do so, the model simulates prices behavior, as described in the following section "6.2.1.1. Fixed Inputs". Then, it multiplies such monthly, stochastically-determined prices by the share of the expected generation determined on the portfolio distribution of the simulation. This result is the monthly revenue from sales, which is done based on the expected generation (Physical Warranty). Next, the model calculates the lack/excess of generation, by subtracting the expected generation from the actual generation. This value, expressed in MWh, is the amount to be liquidated on the clearing market. If the plant is part of the MRE, this liquidation occurs by paying, in case of lack of generation, or receiving, in case of excess, revenues of the TEO price for each MWh cleared. If the plant is out of the MRE, the same occurs, with the exception that the clearing price utilized is the PLD and not the TEO. The sum of the revenues from the Physical Warranty and the Revenues/Losses from the clearing, leads to the final revenues of the plant for each month of the 30-year period.

Lastly, the third step sums the present value of the revenues of each of the months, by utilizing the discount rate required by the investor. In all simulations of this thesis, the discount rate considered was of 20% in real terms. This present value sum is the final output of the model, discussed ahead in terms of expected value (i.e. mean of 1,000 iterations) and standard deviation for each of the 96 simulations.

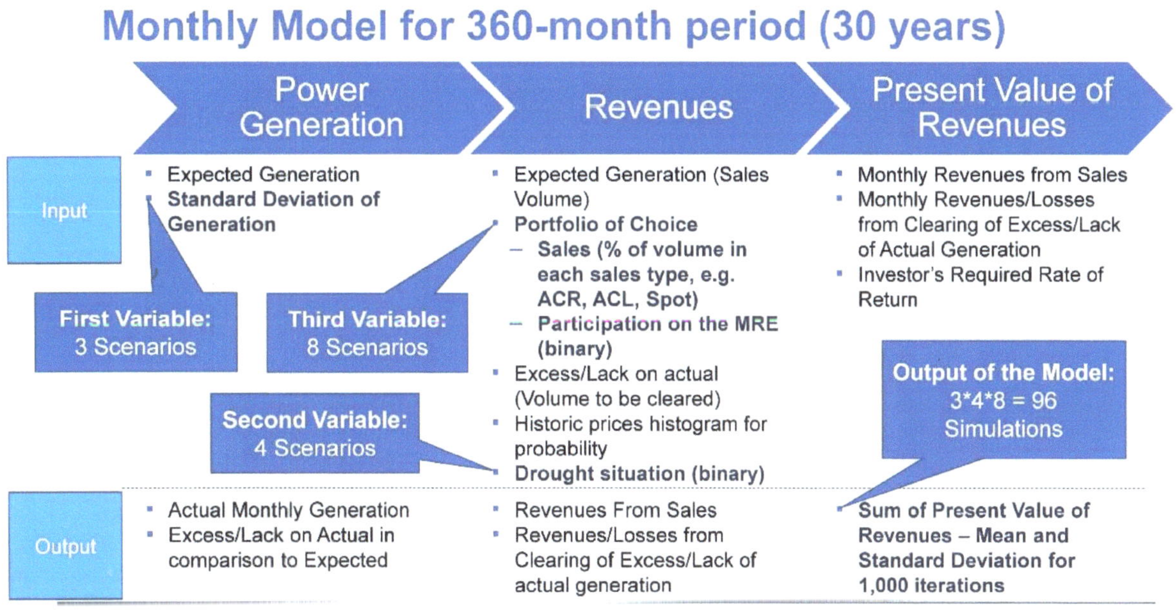

Monthly Model for 360-month period (30 years)

Power Generation	Revenues	Present Value of Revenues

Input
- Expected Generation
- Standard Deviation of Generation

First Variable: 3 Scenarios

Third Variable: 8 Scenarios

Second Variable: 4 Scenarios

- Expected Generation (Sales Volume)
- Portfolio of Choice
  - Sales (% of volume in each sales type, e.g. ACR, ACL, Spot)
  - Participation on the MRE (binary)
- Excess/Lack on actual (Volume to be cleared)
- Historic prices histogram for probability
- Drought situation (binary)

- Monthly Revenues from Sales
- Monthly Revenues/Losses from Clearing of Excess/Lack of Actual Generation
- Investor's Required Rate of Return

Output of the Model: 3*4*8 = 96 Simulations

Output
- Actual Monthly Generation
- Excess/Lack on Actual in comparison to Expected

- Revenues From Sales
- Revenues/Losses from Clearing of Excess/Lack of actual generation

- Sum of Present Value of Revenues – Mean and Standard Deviation for 1,000 iterations

Figure 8 - The Structure of the Model

As mentioned before, the objective of the model created is to understand the impact of three variables in the sum of present value of revenues of a 30-years lifetime operation of a PCH. The three variables are: first, the length of the drought after beginning operations, which is especially useful for PCHs entering the market during the current drought; second, the standard deviation of the monthly generation of plants, i.e. the variation between actual production and the Physical Warranty; and third, the portfolio of energy sales, i.e. the share of the Physical Warranty sold in each of the possible ways (e.g. A-5, A-3, long-term ACL, PLD). Each of these three variables is described in sections 6.2.2. to 6.2.4., and crossing them resulted in 96 different simulations that based the discussion in sections 5.4. onwards. Each simulation was run through 1,000 iterations, thus capturing the impact of the stochastic variables in the output of the model.

The structure of the model within the three steps described above can then be described in four parts. First, the fixed inputs, which are selected once and remain the same for each simulation. Second, the variable inputs, which are different in each simulation. Third, the calculations executed in each iteration, which lead to key data necessary for the final calculation of the output. Fourth, the output itself, which is the sum of the present values of all revenues, after the PLD/TEO clearing is considered.

The fixed inputs of the simulation are divided in three groups: Plant characteristics, Prices Input and Investor's Required Rate of Return. Each of these groups are described as follows.

- **Plant Characteristics**
  - *Installed Capacity:* As this only impacts the absolute amount of the output, and not its proportionality and relations to each of the three variables under scrutiny in this thesis, an arbitrary value of 10 MW have been chosen.
  - *Utilization Rate:* The utilization rate determines the proportion of the annual Physical Warranty and the Installed Capacity. Again, this amount does not affect the proportionality of the output and the variables observed, and thus an arbitrary realistic value of 80% has been chosen. Lastly, it is important to notice that if this model had the objective to understand the return on investment of the plant, this variable would be extremely important and should not be chosen arbitrarily.
  - *Monthly Variation of Physical Warranty:* As described in section "5.3. Installed Capacity and the Physical Warranty of PCHs", the generation in Average Mega-Watts varies monthly, mainly due to the seasonality of water precipitation. In this model, the average production of each month is a fixed input, derived from the analysis of 397 PCHs seen in Graph 36. However, the standard variation of this value is one of the three variable components of the model, and the value actually generated in each month is treated stochastically through the 1,000 iterations of each simulation.

- **Prices Input**
  - *PLD floor and ceiling:* The clearing spot-price PLD currently has values of BRL 30.26 and BRL 388.04 for its floor and ceiling, respectively. This is defined by law, and as the model was built in real terms (i.e. excludes the effects of

inflation), those values were maintained for the entirety of the 30-year lifetime of the plant.

- *PLD variation over time:* The PLD variation depends mainly on the ability of the hydropower plants in the system to supply the energy that defines the marginal cost of the system. This means it depends on maintaining the level of investment in supply compatible to the growth in demand, as well as having similar precipitation characteristics. In the model, it is assumed that the PLD variation in the future will maintain the probability of the past. Thus, the historical data from 2001 to 2015 was used to create a histogram, which is utilized stochastically to determine the values in each iteration. Lastly, this stochastic behavior only take place after the current drought is over, as its duration is one of the variable inputs of the model. During the initial drought period, the PLD is kept at the ceiling, as it currently is and has been through 2014.

- *TEO:* The TEO tariff is review and determined annually by the government. Due to the official calculation method utilized by the government, unless the latter decides to change the regulation, the TEO is expected to have little-to-none variation in real terms. Therefore, it is kept at the current value of BRL 11.25 throughout the 30-years period studied.

- *ACR price within A-5/A-3 auctions*: The price that PCHs sell at the regulated market for a 30-year period prior to construction have varied in the past by growing slower than the inflation. In the 2006-2014 it has increased by a ~3.5% compound annual growth rate, while the inflation was ~5.6% for the same period. However, to analyze PCHs entering the market currently or in the near future, it was chosen to utilize such price from the last auction, of November 2014. The price utilized in the model was BRL 165.26.

- *ACL long-term pricing*: "Determining" the future price of electricity traded with long-term contracts in the free market is not only a merely theoretical exercise, but also the lack of historical data makes even speculation quite difficult. However, the few data points gathered through interviews for this thesis pointed that the hypothesis that such market price follows variations of the spot price is quite plausible. As described in the section 3.4.3., when the PLD is on its low-end, the consumer's cost of waiting for a better deal is not only drastically reduced, but in fact often negative. In this situation, consumers buy on the long-term to reduce the risks of price variation (i.e. "better to pay BRL 100/MWh during 5 years than pay BRL 30 today and possibly BRL 388 next month"). On the other hand, when the PLD is on its high-end, the option of waiting for a better long-term deal becomes very costly for consumers, and selling long-term only becomes interesting for power suppliers if done at a much higher price than during the PLD's low-end. Having said this, the variation of the ACL long-term price utilizes the same PLD histogram than the model, except that it is fit for an interval between the floor of BRL 127.55 and the ceiling of BRL 240.00. It is understood that those two prices lack proper data analysis, but they are based on real data provided by energy investors on interviews. Sadly, the CCEE is still unable to release historical data on bilateral contracts, making the market more inefficient and this part of the thesis less thoroughly data-based than the author would like. Lastly, the term used for these ACL sales contracts were always 2 years.

- **Investor's Required Rate of Return:**
  - *Required Rate of Annual Return:* As it is widely known, Brazilian interest rates are absurdly high, both for companies and for people. In a country where credit card debt sometimes incurs rates of above 300% per annum, and where the average personal interest rate is 115.32% p.a., it would be surprising to see PCH investors expecting 5-10% p.a. rates. In fact, the average company interest rate for borrowing was at 55.19% p.a. as of February 2015 (G1 News; ANEFAC, 2015). However, our interviews showed that PCH investors have a considerably lower than average cost of capital. In

nominal terms, interviews revealed values on the 25-35% p.a. range being used as discount rates to evaluate Small Hydropower projects. With current expected inflation at around 8%, this led the thesis model to have the value of 20% as the real discount rate utilized.

- *Note on the impact of a high Weighted Average Cost of Capital:* The high cost of capital of Brazilian PCH investors direct them towards being "short-sighted". The higher the WACC, the more important the near future revenues become in comparison to the longer term ones. This has major impacts in periods such as the current drought: investors entering the market right now tend to sell much more on short-term options like the PLD, capturing high revenues today in despite of securing long-term contracts. This is understandable when observing a table like the one below. The analysis is simple: consider an equal monthly income for each of the 360 months of a 30-years period. Table 4 shows the share of the present value of the sum of such income within the first 1, 5 and 10 years in comparison to the present value of the total income in the 30-year lifetime. With a Required Rate of Annual Return of 20%, the income of the first 5 years represent 60% of the total present value of the 30-years period. Now, consider the PCH investors who entered the market with a new plant in March 2015 and decided to sell 100% of its energy on the PLD. As they said, they expect the current drought to last 3-5 years at least. If they are right, the present value of this inflated price for this initial period is likely to represent the vast majority of the whole value generated by their PCH. Thus, it is reasonable (yet not necessarily optimal) to decide to have a portfolio fully exposed on the short-term. Lastly, whether this discount rate is optimal for investors, and whether the methods utilized by them in determining such rates is theoretically correct, is a very interesting matter and could have a whole thesis dedicated for it. Yet, this thesis just assumes this value as given.

Table 4 - Impact of the Required Return Rate (Discount Rate) on the share of present value in initial periods

Discount Rate	Discount Rate Monthly	Share of period's present value within 30-year lifetime		
		1-year	5-year	10-year
5%	0.41%	6%	28%	50%
10%	0.80%	10%	40%	65%
15%	1.17%	13%	51%	76%
20%	1.53%	17%	60%	84%
25%	1.88%	20%	67%	89%
30%	2.21%	23%	73%	93%

6.2.1.2.   Variable Inputs

The objective behind the three variable inputs is to test their impact into the output and its standard deviation. As detailed in sections 6.2.2. to 6.2.4., each of these variables had values chosen based on historical data available, and each combination of their settings was tested, leading to a total of 96 simulations. The first variable, the standard deviation of the actual monthly generation, has three different scenarios. The second variable, the length of the current drought, has four different scenarios. Lastly, there are 8 portfolio combinations of sales considered. Thus, there were 3 x 4 x 8 = 96 simulations ran, with 1,000 iterations each to allow the capturing of the stochastic behaviors.

6.2.1.3.   Calculations per Iteration

Each iteration generates values based on the stochastic behaviors described and the fixed inputs, calculating a series of data over the 360-month period. This data is later used to calculate the actual output of each iteration, and the 1,000 iterations of each simulation allowed to understand the statistical behavior of the output. Such calculations are as follows.

- **Monthly Actual Generation** – the model calculates, in each iteration, the amount of energy generated each month. It does so by generating data stochastically, based on a normal curve with average matching the Physical Warranty. Therefore, the model considers that over time, the expected generation of the plant is the

Physical Warranty itself, which matches the method utilized by the government to create such concept.

- **Electricity Prices** – As described in section "6.2.1.1. Fixed Inputs", prices vary depending on the length of the current drought, and after that based on a histogram created from historical data of 2001-2015. Therefore, for each iteration the model generates a monthly value of prices, based on such stochastic behavior.

- **Lack/Excess of Generation liquidated via PLD or TEO** – Depending on the results of the Monthly Actual Generation, it is likely to exist either an excess or a lack of production in comparison to the Physical Warranty. If the PCH is part of the MRE pool – a decision that is part of the portfolio – this liquidation will happen via the TEO tariff. If the PCH is not part of the MRE, it happens via the PLD. In any case, excess of production represents extra revenue, and lack of production represents the need to purchase the energy to complete its offering, thus representing losses.

- **Sum of monthly revenues/losses** – In each iteration, each month's final revenue is calculated by summing the revenues from the Physical Warranty-based sales (i.e. revenues from the portfolio that sums 100% of Physical Warranty), to the extra revenues or losses of the liquidation of lack or excess of generation described above.

6.2.1.4. The Output

The output of the model is then calculated by summing the present value of the revenue of each of the 360 months, given the Weighted Average Cost of Capital of the plant's investors. This is done in each iteration, and for each simulation an "output curve" is built with 1,000 iterations. The objective of the model is to understand the average and standard deviation of this output curve for each of the 96 simulations, thus understanding the impact of the different combinations of the three variables in the expected present value of revenues of the PCH.

6.2.2. First Simulation Variable: The Standard Deviation of Generation

The first variable is the standard deviation of the actual monthly generation, thus its difference from the Physical Warranty. This deviation is measured in percentage of the Physical Warranty. The actual monthly generation varies with the Physical Warranty (100%) being the average, and this first variable indicated how stretched the normal curve around it should be. The model picks "random" values within this stochastic distribution for each iteration, and over the 1,000 iterations of each simulation it is possible to observe the statistical impact of this variation on the output.

As mentioned before, the average production of each month of the year was defined based on the analysis of the 397 plants available at CCEE's databases. The standard deviation, however, can be arbitrarily chosen based on several different scenarios. Seven different methods were used to estimate each month's standard deviation based on the 397 plants database, over the 2001-2014 period. The first three methods were done for the total group and for a group excluding outliers, i.e. excluding plants with production lower than 50% or higher than 130% of the Physical Warranty. Such methods consist in comparing each month's average per plant, each month's average by all plants available and each month's average per plant normalized by the average production (thus assuming that the Physical Warranty might have been miscalculated by ANEEL). The last method was simply by observation of benchmarks.

Although each plant's deviation seems to be consistent over time, each plant may have very distinct variation profiles. Due to this, three variation profiles were chosen to be used across simulations: one with each month's deviation on the ~5% vicinity, one on the 15-20% and another on the 25-30%. These three extremes are demonstrated in Graph 39, and support the observation of the impact on the deviation of the output cause by the discussed variable.

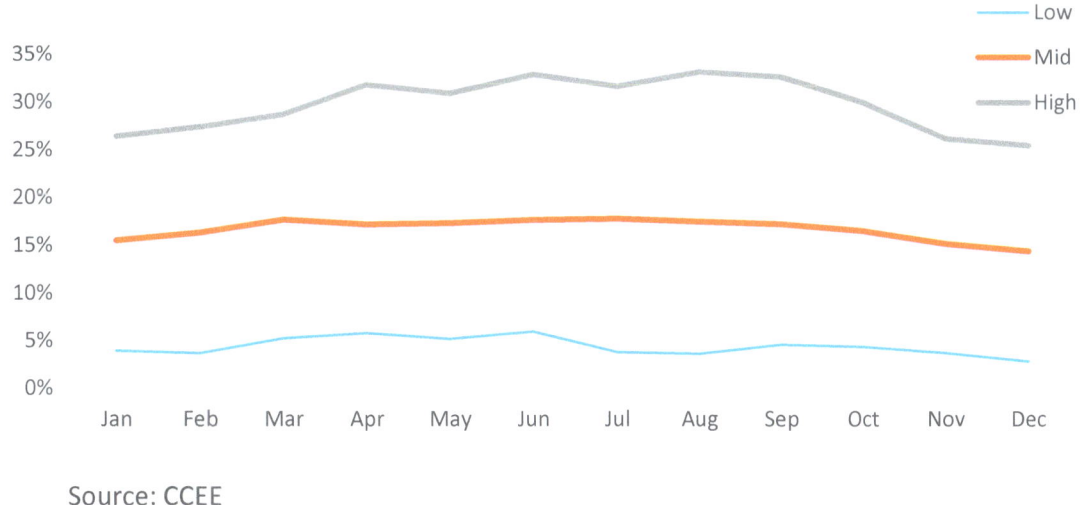

MONTHLY GENERATION STANDARD DEVIATION

Low, Mid and High Standard Deviation Profiles used on Simulations

Source: CCEE

Graph 39 - First Model Variable - Three Monthly Generation Standard Deviation Scenarios

6.2.3. Second Simulation Variable: The Length of the Continuing Drought

The second simulation variable is the length of the continuation of the drought that exists at the moment that the plant starts to operate. As seen during interviews, investors of PCHs entering the market today believe that the drought will continue for 3-5 years. If they are right, investors that begin the process of building a PCH today may enter after the drought is over and the normal "modus operandi" of the market is back on track. As the drought has a very large impact on the price of the PLD, making it be constantly at its ceiling, and as this is happening in the first years of operation, it has tremendous impact on the output. Thus, four "current drought" scenarios have been utilized in the simulation: One where the plant starts to operate after the drought is over, and three scenarios where it lasts 1, 3 and 5 years respectively. In all scenarios, within the drought the PLD is always at the ceiling, and outside the drought it obeys a probability derived from the behavior of the PLD during the 2001-2015 period. This was constructed through the utilization of a histogram of PLD values, considering historic values at BRL of January 2015.

6.2.4. Third Simulation Variable: The Portfolio Choices

The last variable, or more specifically "set of variables", for the model is the portfolio of choices that impact sales. This set of decisions are constructed as follows. Each variable within this portfolio has two instances: one during the drought period, and one for after the drought is over. For instance, an investor that choose to sell 100% of its Physical Warranty in the spot price during the drought may change to another portfolio strategy once he or she realizes the drought is over and the PLD starts to decrease.

- **Participation on MRE** – Each year, the plant can decide whether to join or not the MRE, thus whether lack or excess of production will be traded at PLD or at TEO. For the sake of simplicity, the model emulates this annual decision in only two moments: joining or not the MRE during the drought, and joining or not the MRE after the drought is over and the PLD behaves similarly to the 2001-2015 history. Thus, this part is constructed in model through two binary values.

- **Share of Sales at PLD** – although the PLD is a clearing price, PCHs may choose to sell energy in the spot market, either through PLD or a very short-term (i.e. monthly) ACL contract. As mentioned before in this thesis, PCHs plan the sales for 100% of the Physical Warranty, while any variation in generation from that is cleared to PLD or TEO. Thus, out of this total, any share between 0 and 100% can be sold at the spot market. As mentioned, this value is input in model for during the drought and post-drought.

- **Share of Sales at the Long-Term ACL** – Similarly to the PLD described above, any share between 0 and 100% can be sold at the ACL. In the model, this is also constructed in two percentage values, one during the drought and one post-drought.

- **Share of Sales at the Long-Term ACR** – A-5 or A-3 auctions, as described in section 3.4.1., allow PCHs to sell energy 5 or 3 years prior to operations, respectively, for a 30-year period. This amount of energy can be anywhere between 0 and 100% of the Physical Warranty. However, differently than in the other sales methods described above, the PCH cannot change this share after it sold on the ACR auctions. Thus, this is a decision taken at least 3 years prior to operation that commits a share or the totality of the generation on a sale at a given price, only to be readjusted by inflation.

In conclusion, the 8 portfolio combinations utilized to form the 96 simulations were as follows. They vary from 100% sold at the 30-year ACR option to 100% sold at the PLD spot market. The combinations, shown in Table 5, were arbitrarily chosen based on interviews and data on PCH investors' behavior.

Table 5 - Portfolio choices analyzed through the 96 simulations of the model

Portfolio Type	MRE Drought	MRE Post-Drought	PLD Drought	PLD Post-Drought	Long-term ACL Drought	Long-term ACL Post-Drought	ACR A-5
1	0	0	100%	100%	0%	0%	0%
2	0	1	100%	0%	0%	100%	0%
3	0	0	70%	20%	0%	50%	30%
4	0	1	70%	0%	0%	70%	30%
5	1	1	0%	0%	0%	0%	100%
6	1	1	0%	0%	100%	100%	0%
7	0	0	30%	30%	0%	0%	70%
8	0	1	30%	0%	0%	30%	70%

6.3. Possible Revenue Sources: The Decision Flow and its Timeline

In section "4.4. The Energy Trading in Brazil" there is a detailed explanation on the regulated (ACR) and the free (ACL) markets of electricity in Brazil. In this section, it is important to understand how the timeline of decisions of sales take place in the lifetime of a PCH. An overview of these decisions can be found in Figure 9. The ACR, due to its regulated nature, have a much more formal and pre-determined timeframe for the PCH investors, while the ACL provides a significant flexibility due to the freedom of creating bilateral deals it provides.

The first sales decisions the PCH investors have to make happen long before the plant is operational. PCHs may join in A-5 or A-3 auctions, which allows them to sell any share of their future production (i.e. share of the physical warranty) for a period of 30 years. Such auctions happen 5 and 3 years, respectively, prior to the beginning of the actual generation by the PCH, thus allowing investors to commit a part or the totality of their production for a price determined at the date of the auction, only to be readjusted by the inflation. Maximizing the share of A-5 or A-3 energy sold in the portfolio of the PCH significantly reduces the standard variation of the future revenues of the plant. In combination with the participation in the MRE pool, this allows investors to have very high certainty over the future revenues. However, reducing such risks comes at the expense of selling at a price that could potentially be lower than if selling on shorter term periods.

Still on the regulated market, there is the annual option of selling at A-1 auctions. Each year, ANEEL and CCEE host auctions called "Energia Existente", which allows currently operational plants to sell in the regulated market as well. However, the term is significantly shorter than at A-5 or A-3 auctions. Only three PCHs, out of the 448 operating ones, have sold in A-1 auctions prior to the writing of this thesis. They have done so for periods of 5 years, at a price considerably higher than the longer-term auctions: BRL 270 in mid-2014, compared to BRL 137 of A-3/A-5 PCHs in December of 2013.

Another very relevant annual decision for the sales portfolio is whether or not the PCH will be part of the MRE. As demonstrated in section "5.4. The MRE: An Energy Relocation Mechanism to Reduce Risk", this can cause very significant impact in the expected value of the sales of a PCH. This impact can become especially heavy, with potential for enormous losses, in case the PCH cannot generate energy to the level of its physical warranty for an extended period.

As for the ACL, decisions are much more flexible and with little pre-determination of time. A bilateral agreement can be closed at any point of the lifetime of a Small Hydropower Plant. It is done in secrecy, between the two parts only and registered at the CCEE while undisclosed to the public. The period covered in such contracts can be anywhere from a month to several years. For simplification, in the model described in the following section, a standard period of 2 years was utilized for ACL contracts. For short-term bilateral contracts, the model utilizes the PLD itself (i.e. assuming a 1-month contract with a 0 spread over the PLD). Observing measurements such as the "Índice Spot" of the company BBCE (BBCE, 2015) shows that this assumption is fair and close to reality.

Finally, the PCH investor may also choose to sell the generated energy only on the market clearing. In other words, this means that the plant provided energy for the market but did not sell it previously; thus, there is someone that utilized more than purchased and will be required to pay PLD price for that difference. In this case, the investor is not participating in the MRE, as if it were it would be selling at TEO instead of PLD, meaning a very low price always.

With all this options together, the investor will always sell 100% of its production – which makes sense when considering the nature of electricity, only delivered when consumed. The expectation of such production is the Physical Warranty, which is the amount that guides (as well as limits) the investor's commitment of sales in advance. After the month is passed, the actual generation may have been higher or lower than the Physical Warranty, thus being cleared via PLD or TEO, depending on the participation on the MRE.

Decisions of the Portfolio of Sales

Period	Decision	Description
Once	Sale at A-5 Sale at A-3	• A-5 and A-3 stand for 5 and 3 years before the powerplant starts generating • Plants can sell up to the Physical Warranty through auctions for up to 30 years • Last price levels were between BRL 130 and 170 (Regulated market ACR)
Annual	Sale at A-1 Entry in MRE	• A-1 allows plants to sell up to the total Physical Warranty through auctions • Periods of up to 8 years, usually 5 • Last price levels were the same as above, but ceilings were BRL 201 and 180 for 3 and 5 years respectively • The MRE is a pool of energy – those who produce more than the Physical Warranty commit to sell at a very low price to those who produce less, mitigating the exposure to the PLD in case of regional droughts • The decision to participate in the MRE is annual and follow a strict schedule
Short-term	Sale at PLD Sale at Short-term ACL	• The PLD is a price regulated by the government – **AFTER** the period, plants that sold more than produced buy at PLD, and vice-versa • The short-term ACL (free market) is the purchase of monthly power **PRIOR** to the period, and usually is comprised of PLD + or - a small premium
Long-ter	Sale at Long-term ACL	• The long-term ACL (free market) is unregulated and so far has no disclosure • Prices tend to be lower the longer the term of the contract is • Usually transfer energy from 6 months to up to 5 years

Source: CCEE (Chamber of Commerce of Electrical Energy); ANEEL; Brazilian Laws; Interviews

Figure 9 - The Decisions that Constitute the Portfolio

6.3.1. The Correlation of Shorter-term Auctions to the PLD

Although there is a complete lack of data on bilateral trading within the free market, there significant amounts of data on auctions. Plants can sell electricity for periods of up to 30 years through auctions, but contracts that cover periods of 5 years or less can be used as a proxy of the free market price – not that their levels would be equivalent, but their variation should follow very similar patterns. In fact, reduced taxes and tariffs allow generators to sell for higher prices in the ACL than in the ACR. Although this differences offset prices, they should not alter the high correlation of variation among prices practiced in the ACR and the ACL.

Graph 40 compares the average prices practiced by generators in auctions while selling electricity in contracts with periods of supply of 5 years or less. It corroborates the hypothesis that ACL contracts should act significantly correlated to the PLD prices. Moreover, as observed in spot prices such as the ones declared by the company BBCE (BBCE, 2015), the shorter the term of the bilateral agreement, the closer it gets to the PLD. Indeed, as theory shows, the future prices converge upon the spot price as the deliver month approaches. The famous didactic website Investopedia has a great section on this topic, for those who are not familiar with derivatives[xvii].

[xvii] This section ca be found at http://www.investopedia.com/ask/answers/06/futuresconvergespot.asp

SHORT-TERM AUCTIONS VS. PLD

Average auction price for sales of 5 years or less of supply versus average annual PLD in nominal BRL

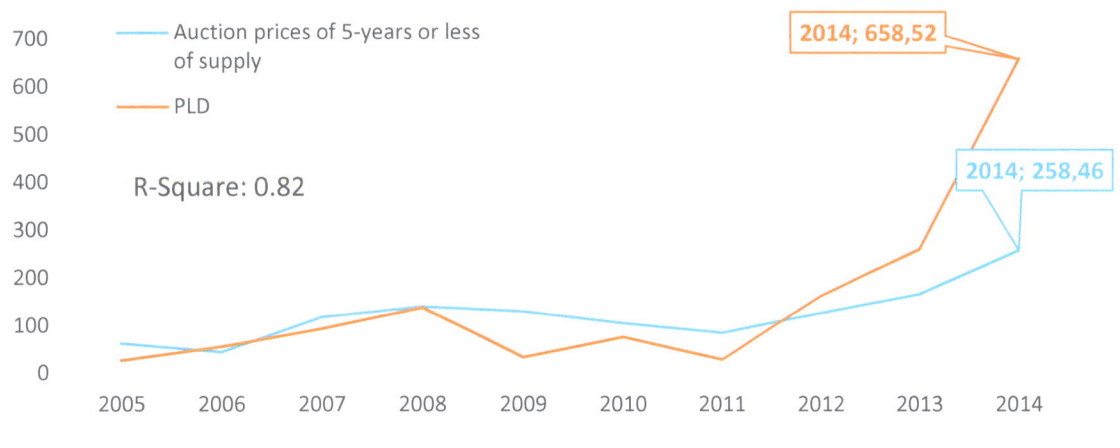

Source: ANEEL; CCEE

Graph 40 - Short-term Auctions Prices versus Annual Average of PLD in Nominal Terms

6.3.2. A-5/A-3 Prices for PCHs in Real Terms

An incautious investor could assume that the A-5/A-3, 30-years sales prices should remain approximately equal year-over-year in real terms. Indeed, after its auction, the plant will receive the same price agreed for the full extent of the 30 years, only to be adjusted by inflation. However, long-term auctioned prices vary with a number of factors – e.g. the PLD behavior at time and prior to the auction, the supply/demand balance within that given auction, the ability to navigate bilateral sales of the investors in the auction and possibly other factors more. Having said that, it is out of the scope of this thesis to understand each of those variables, although it could be a great field for further study.

Graph 41 shows the average prices sold by Small Hydropower Plants for 30-year periods in auctions. Although there is data available in only 6 years on the 2006-2014 period, the data seems to point a decline in prices over time. Prices practiced in 2014 were about 14% lower than the ones practiced in 2006, and over said period there seem to be a trend of overall decline. In any way, by the time the investors join A-5/A-3 auctions, they would not have begun to build their plants yet. Thus, there is value in the option of postponing the investment, especially should the price of the auction not meet their expectations. This value makes this sales option very interesting, as it allows investors to gain a tremendous reduction in risk of future revenues of their plants, while still being able to observe and act on prices prior to the full commitment of the construction.

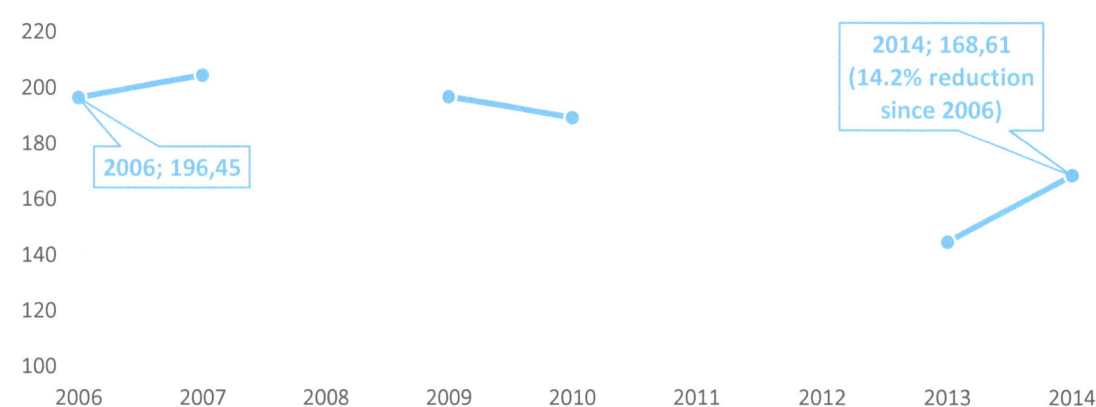

Graph 41 - PCH Prices at Auctions in the 2006-2014 Period, in Real Terms

6.4.   Understanding the Output Results Across Simulations

The results are presented following in the form of 3 pair of tables. Each pair is derived from one of the three types of monthly generation standard deviation (Low, Medium and High). A lower standard deviation means that the plants actual production is closer to the Physical Warranty, i.e. the monthly actual generation normal curve is shorter on the X axis. Furthermore, each pair consist of a table of the average of the output (i.e. the expected value of the sum of the present value of revenues) and a table of its standard deviation (i.e. a measure of risk of the output). All tables are organized in Portfolio types (1 to 8, as described in section "6.2.4. Third Simulation Variable: The Portfolio Choices") versus length of the drought currently undergoing when the plant starts to operate (0, 1, 3 or 5 years, as described in section "6.2.3. Second Simulation Variable: The Length of the Continuing Drought").

The output data is presented next in this section, in Table 6. Insights derived from the 3 pairs of tables are discussed in the following sections.

Table 6 - Output Average and Standard Deviation from the Thesis Model

PCH with Low Standard Deviation of Monthly Generation

	Output Average						Output Standard Deviation			
	Drought Years						Drought Years			
Portfolio Type	0	1	3	5		Portfolio Type	0	1	3	5
1	52,9	68,7	93,0	109,9		1	4,3	3,6	2,6	1,8
2	61,6	76,0	98,0	113,4		2	6,4	5,4	3,8	2,6
3	60,2	70,6	86,5	97,5		3	3,6	3,0	2,1	1,5
4	62,0	72,1	87,5	98,2		4	4,5	3,8	2,7	1,9
5	62,9	62,9	62,9	62,9		5	0,0	0,0	0,0	0,0
6	61,6	70,7	77,1	81,5		6	6,4	4,5	3,0	2,1
7	59,9	64,7	72,0	77,0		7	1,3	1,1	0,9	0,8
8	62,5	66,9	73,5	78,1		8	1,9	1,6	1,2	0,9

PCH with Medium Standard Deviation of Monthly Generation

	Output Average						Output Standard Deviation			
	Drought Years						Drought Years			
Portfolio Type	0	1	3	5		Portfolio Type	0	1	3	5
1	52,9	68,7	93,0	109,9		1	4,4	3,9	3,2	2,7
2	61,6	76,0	98,0	113,4		2	6,4	5,5	4,1	3,2
3	60,3	70,6	86,5	97,5		3	3,7	3,3	2,8	2,5
4	62,0	72,1	87,5	98,2		4	4,5	4,0	3,2	2,7
5	62,9	62,9	62,9	62,9		5	0,1	0,1	0,1	0,1
6	61,6	70,7	77,1	81,5		6	6,4	4,5	3,0	2,1
7	59,9	64,7	71,9	77,0		7	1,7	1,8	2,1	2,1
8	62,5	66,9	73,5	78,1		8	1,9	2,0	2,1	2,1

PCH with High Standard Deviation of Monthly Generation

	Output Average						Output Standard Deviation			
	Drought Years						Drought Years			
Portfolio Type	0	1	3	5		Portfolio Type	0	1	3	5
1	52,9	68,7	93,0	109,9		1	4,7	4,5	4,3	4,1
2	61,6	76,0	98,0	113,4		2	6,4	5,8	4,9	4,3
3	60,3	70,6	86,5	97,5		3	4,0	3,9	4,0	3,9
4	62,0	72,1	87,5	98,2		4	4,5	4,4	4,1	4,0
5	62,9	62,9	62,9	62,9		5	0,1	0,1	0,1	0,1
6	61,6	70,7	77,1	81,5		6	6,4	4,5	3,0	2,1
7	59,9	64,7	71,9	77,0		7	2,3	2,8	3,5	3,7
8	62,5	66,9	73,5	78,1		8	1,9	2,7	3,4	3,7

6.5.   The Expected Present Value of 30-Year of Revenues

The first interesting insight comes from simply observing the shape of the curves around the output for each portfolio combination, while disregarding the effects of the drought. Graph 42 shows such curves for each portfolio, considering the medium scenario of standard deviation of monthly generation.

This observation begins at portfolio number 5, which shows a curve with very high certainty on the present value of revenues for the 30-year period. Indeed, as seen in section "6.2.4. Third Simulation Variable: The Portfolio Choices", such portfolio consists of selling 100% of the Physical Warranty at the A-5/A-3 auctions, while participating during the whole period in the MRE pool. This not only provides a very high certainty on the revenues, but also provides the best expected value given that the PLD behaves in line with the histogram generated by the 2001-2015 period.

However, as it can be seen in other drought scenarios, this certainty does not allow investors to capture the value of periods of scarcity of supply. As observed in the results of section "6.4. Understanding the Output Results Across Simulations", an investor that enters the market and faces a drought during the first year of operations would already have portfolio number 5 shifted to be the lowest expected (i.e. mean) output of all 8 combinations. This is due to the tremendous impact in present value of selling at very high prices during the first year – as it would happen by selling at the drought's PLD of BRL 388.04, as opposed to an A-5/A-3 price of around BRL 170.

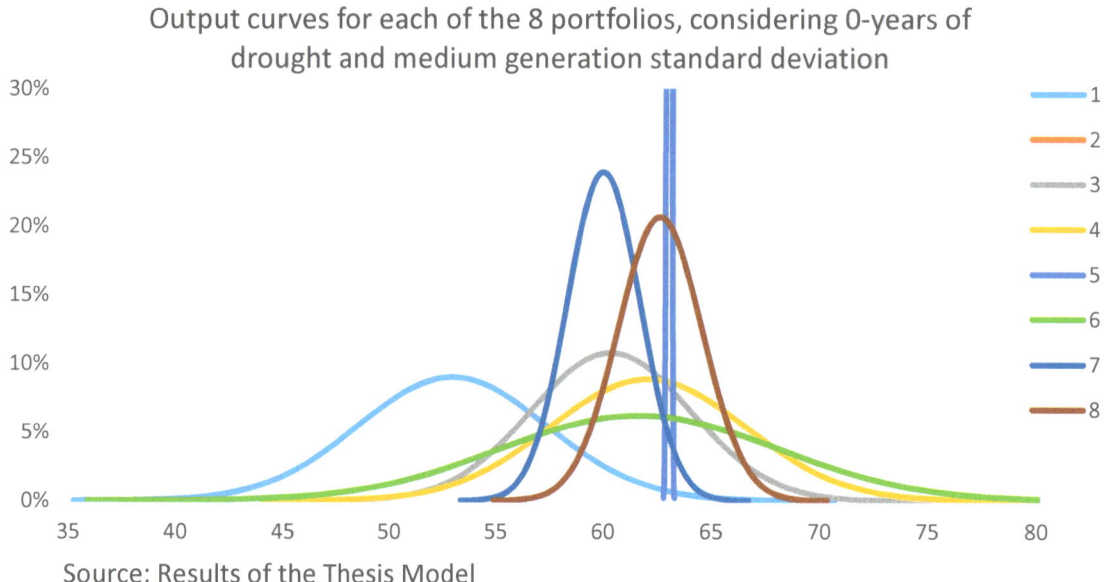

OUTPUT CURVES PER PORTFOLIO

Output curves for each of the 8 portfolios, considering 0-years of drought and medium generation standard deviation

Source: Results of the Thesis Model

Graph 42 - Output Curves for Portfolio Choices Considering Plants Entering Post-Drought

6.6.   The Impact of Imprecise Monthly Generation

As expected, the increase in the standard deviation of monthly generation did not impact the average of the output. However, it is important to notice that this only happened because in all three scenarios of this variable the average monthly production was kept equal to the Physical Warranty. In theory, this should be the long-term reality; otherwise, ANEEL would revise the Physical Warranty of the plant to lower values. Having said that, an underperforming plant that has a lower average than the Physical Warranty will surely have reduced revenues, and although this variability is unimportant in the context of this thesis, it should be considered by PCH investors.

In terms of the standard deviation of the output, however, this variable has proven to be impactful for non-MRE participants. Portfolios 1, 3 and 7 have no participation on the MRE throughout the lifetime of the plant, and they present high increases in Standard Deviation of the output, as shown in Graph 43. Furthermore, by analyzing Graph 43 it is possible to see that in periods of drought, the lack of protection against the PLD fluctuations (i.e. not participating in the MRE) greatly increases the impact of the variation in actual monthly production. On the other hand, in periods with PLD variation close to the 2001-2015 history, the variation of actual generation does not have a tremendous impact on the deviation of the output.

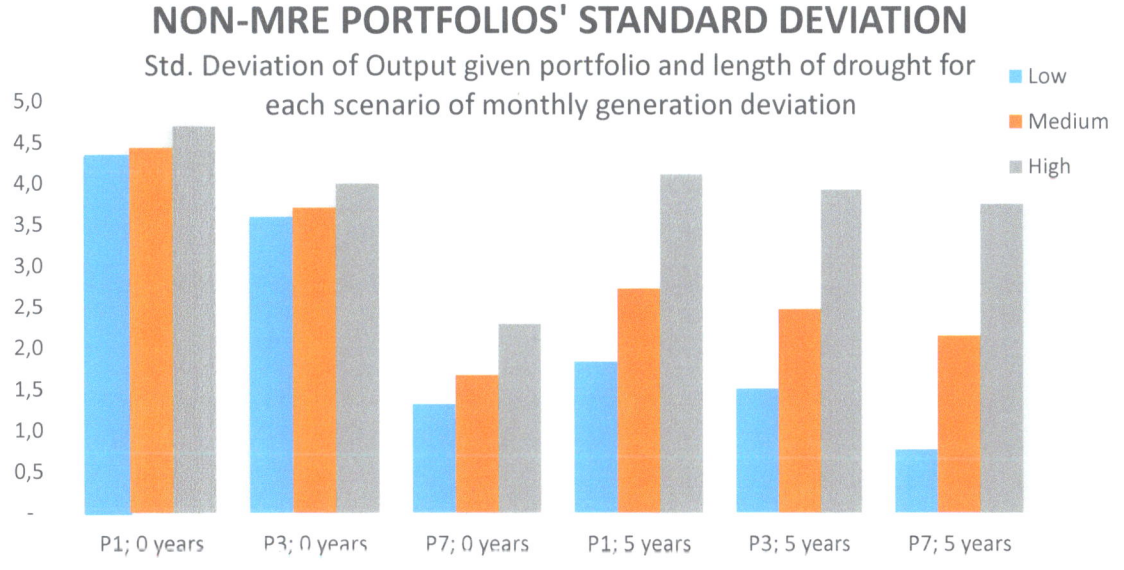

Source: Results of the Thesis Model

Graph 43 - Standard Deviation of Output in Non-MRE Portfolios

6.7.   The Current Drought and its Effects on Revenues

As explained in section "6.2.1.1. Fixed Inputs", due to the high WACC of Brazilian investors, the majority of the present value of PCHs, especially those exposed to the PLD, was expected to reside in the first 5 years. Graph 44 shows the impact of such exposition in terms of length of the initial drought. It is important to note that no portfolio choices have a decrease in expected output, as they always either have at least 30% sales dedicated to the PLD or are part of the MRE. Although unrealistic, a portfolio 100% sold out of the spot market, e.g. 100% sold at the ACR, would suffer severely in terms of standard deviation of revenues if not part of the MRE. Finally, portfolios 1 and 2 have 100% dedicated to the PLD during the drought, while 3 and 4 have 70% on the same period, and the pairs 5/6 and 7/8 had 0% and 30% respectively.

Portfolio 5 is worth mentioning, as it is the only one that does not show an increase in output with the drought, which is expected since it is 100% dedicated to the 30-year ACR sale. Moreover, the participation in the MRE throughout the whole period also makes it standard deviation of output significantly lower than any other portfolio. In other words, Portfolio 5 is the "safe play" – even before construction, investors know with a quite high degree of confidence how much they will have in revenues for their PCH. Worth mentioning, this is considering a healthy operation of the PCH (i.e. excludes malfunctions and defects that may lead to reduction of physical warranty).

Another interesting insight is that the standard deviation of the output in portfolios with high exposure to PLD or the ACL, namely 1, 2, 3, 4 and 6, is reduced with the increased length of the drought. This is because the drought causes the PLD and the ACL to be constantly higher. In other words, the marginal cost of generation of the national system, which defines the PLD, is always above the ceiling of BRL 388.04 – thus the value of the PLD is kept constantly at this value.

Lastly and most importantly, while it is true that scenarios with more PLD/ACL combinations have higher standard deviation of output, as shown in Graph 45, their expectation in general is higher, as shown in Graph 46. In fact, the justification heard from investors of leaving the portfolio fully exposed to the PLD, while planning to retreat back to the ACL after the drought is over, seems to make sense. Not that investors are able to foresee when the drought will be over – actually, this forecast exercise is irrelevant – but when the PLD fall from the currently stagnated position at its ceiling, investors will be able to move to sell at longer terms, lower prices ACL contracts. Within such movement, they have the option to join the MRE, allowing them to do the transition described in portfolio number 2. This transition would allow them to capture significant upside from the current situation, while still maintaining a healthy expected output even if the drought does not last as much as they expected.

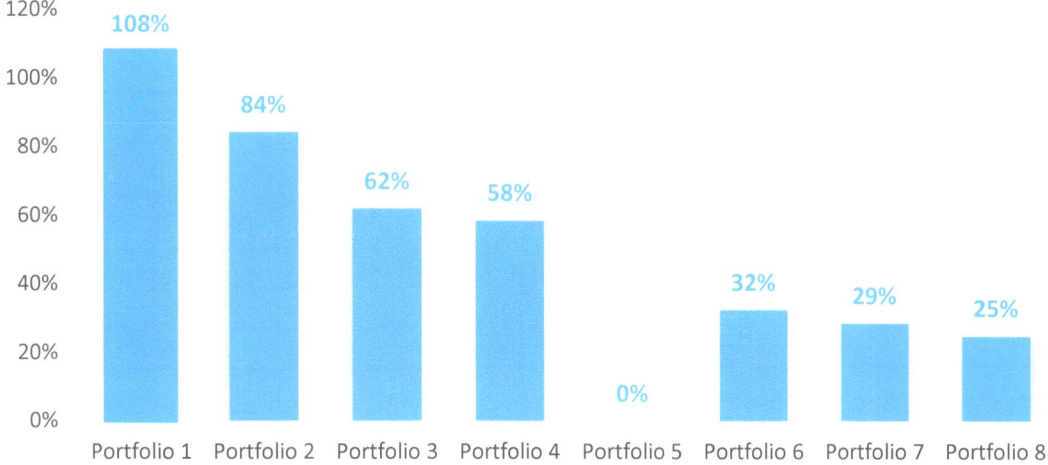

Source: Results of the Thesis Model

Graph 44 - Increase in Output due to 5-years of Drought in the Beginning of Operations

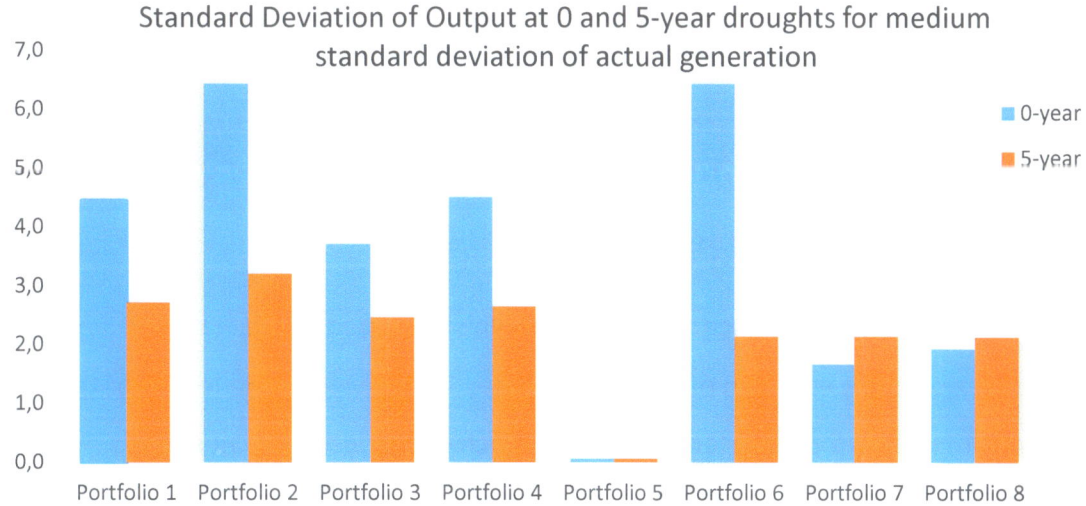

Source: Results of the Thesis Model

Graph 45 - Increase in Standard Deviation of Output due to Drought

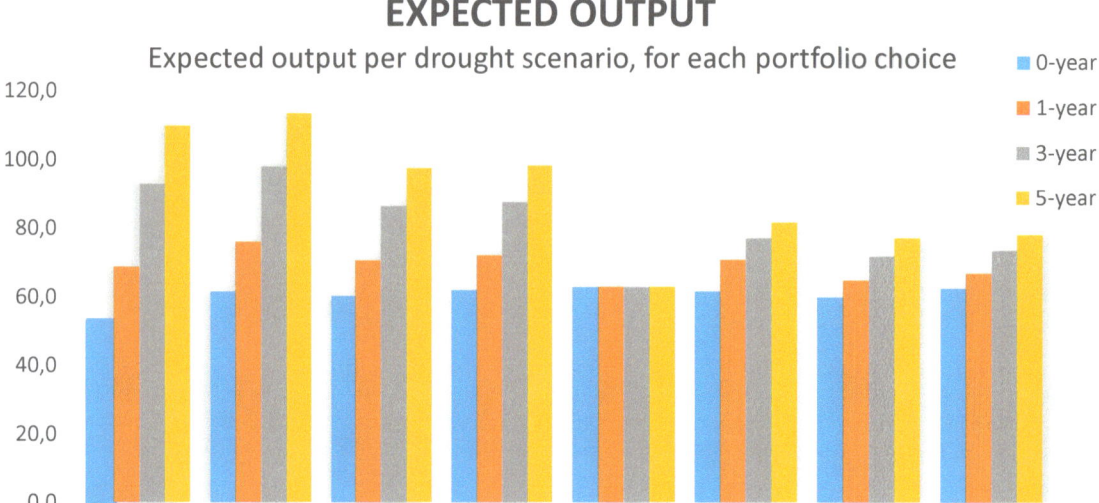

EXPECTED OUTPUT

Expected output per drought scenario, for each portfolio choice

Source: Results of the Thesis Model

Graph 46 - Variation in Expected Output of Portfolios in Each Drought Scenario

6.8. Understanding Risk in Small Hydropower Investments

Observing the results of the model in terms of the output and its standard deviation shows that the net present value of the investment in a Small Hydropower facility can vary significantly. The three variables observed across the simulations – deviation in actual generation, length of the current drought and the portfolio decisions – have each one of them different impacts on the risk and its perception by investors.

It is also important to notice that risk is not only observed in the standard deviation of the output, but also on the variation of the mean caused by the increase in length of the drought. A portfolio more exposed to shorter-term prices, such as the portfolio number 1, is much more exposed to the price volatility. In this case, the drought was the cause of such variation, and even while maintaining the current floor and ceiling values of the PLD over time, this variation could cause more than 100% variation in the expected present value of revenues. Although not considered in the model, other causes may have similar effects. As an example, until the end of 2014, the ceiling of the PLD was BRL 822.23/MWh. In November of 2014, the PLD was traded at this maximum allowed price. Two months later, the government slashed this ceiling to less than half, leading to its current value of BRL 388.38/MWh. This situation was completely unexpected

106

until a few months before it actually happened, meaning that investors are strongly exposed to the risk of regulatory changes.

On the other hand, the overall regulatory structure of the Brazilian electricity market has been very stable since its implementation in 2003. While the government made a few changes, such as increase in tariffs and decrease of the PLD ceiling, it maintained the structure of auctions, trading and investment freedom. Furthermore, there were only positive signs in terms of the government's inclination towards Small Hydropower investment. As described in the section 3.5., this electricity source is seen as better than most others in many dimensions, from being relatively fast to build to having very low environmental impacts. Furthermore, options such as the long-term auction A-5/A-3 sales also enable investors with smaller risk appetites to play in this market.

Another important risk to be considered by investors is the difficulty of forecast when it comes to actual generation of Small Hydropower Plants. The long-term behavior of water flows and precipitation is not only a very complex science, but it is also often far from the realm of knowledge of energy investors. Predicting major droughts, such as the one that currently takes place in the southeast of Brazil, is nearly impossible when considering a 30-year timeframe. Furthermore, Small Hydropower Plants are often built as "Run-of-the-River", meaning that they have little or no reservoirs. Those plants are much more exposed to variations in water flows than the large hydropower plants, and a long-term reduction in flow could mean expulsion of the MRE, reduction of Physical Warranty and other effects that can dramatically reduce the present value of revenues.

Often, investors build Small Hydropower Plants as a measure to mitigate the financial risk of their energy-intensive industries. Still, in the event of a drought that affects their generation, they will remain exposed to clearing prices. However, with a Small Hydropower Plant they have the option to be exposed to the lower price of the TEO, by joining the MRE pool, which means that they could greatly reduce exposition to volatility of electricity prices for their industries.

From the Small Hydropower investor, there are tremendous gains in building more than a single plant – reduction in capex, gains in trading, increase in precision of Physical Warranty prediction, as described before. This means that a simple yet powerful risk mitigation measure would be to enter with multiple plants, which would be even more powerful if they are located in very different regions of the country. This not only enables investors to replicate methods and learn by repetition, but also brings gains from scale. As an example, one of the interviewed investors reported that by having invested in 5 plants already, their company became a local reference for smaller investors, who then allow the interviewee to manage the trade of their energy in exchange for a fee (Small Hydropower Investors, 2015).

As final observation on risk, the output of the model shows only 8 possible combinations of portfolio of sales. However, the optimization of the combination of portfolio decisions varies with the investor's capabilities and risk appetite. For instance, an investor with better meteorological data and knowledge may feel more comfortable drought-related risks. On a different perspective, an inexperienced investor may choose to have a portfolio completely sold on A-5/A-3 auctions on his first plant, thus focusing on optimizing capital expenditure and other risks that are significantly higher in a first plant. In conclusion, each investor should build an optimization model, taking into account their specific characteristics and risk perception.

7. Conclusion and Further Studies

The main conclusion of this thesis for those who were already familiar with the Brazilian Electricity Market is that the wide range of portfolio combinations that can be chosen by Small Hydropower Investors may create interesting risk-reward ratios. Furthermore, it also shows that assuming a 30-year period of prices distributions in line to the 2001-2015, i.e. assuming there is no initial drought period, the expected value of a portfolio highly based on long-term A-5/A-3 sales can be higher than more exposed portfolios, while still having insignificant standard deviation. On the other hand, such low-risk portfolio does not allow investors to capture the value of times when the supply is scarce, such as in the current drought. Furthermore, such scarcity periods can cause tremendous impact on the expected return of shorter-term portfolios. Lastly, all these variations in price are significantly more relevant in the first few years of operation, as demonstrated in this thesis through the observation of the high rates of return required by Brazilian investors.

The efforts of this thesis lay ground to Brazilian Electricity Market scholars to create further knowledge that may be directly utilized by investors with the objective of optimizing their investments. This document provides a general overview of the market and some insights on the most important variables for such investors, including the benefits and trade-offs of optimized portfolio construction in electricity sales. However, there is still several fields of study that could benefit the Brazilian Electricity Market, providing tools to make it more efficient.

The first of those topics relates to the actual optimization of the portfolio of electricity sales. This thesis demonstrates the extent of the impact, and thus the relevance of portfolio-thinking on the electricity market. However, it does not provide tools to enable investors to effectively optimize their portfolio based on their risk appetites.

The second topic is regarding the portfolio-thinking in terms of the electricity sources chosen by investors. While this thesis provides some understanding of portfolio-thinking in terms of methods of electricity sales, an analogous work could be done while considering that investors may have portfolios with hydro, wind, thermal, solar and other kinds of power sources. Benefits derived from negatively correlated variations in returns could be found with thorough analysis of these possibilities.

The third topic is to further understand other value creation levers for hydropower investors. As mentioned before, levers that were not explored in this thesis are such as financial leveraging, tax reduction/shields, operational improvements, increases in efficiency of construction techniques or even reduction in construction time. Each one of those could significantly impact the return on investment of hydropower plants – thus, further studying the impact of each one of them, specially within the PCHs context, could be extremely relevant for investors.

The fourth topic is to explore the correlations (and causations) among the different types of electricity sales. The PLD is defined by the marginal cost of operation, but there is no quantitative analyses regarding how the longer-term free market prices are affected by it. Furthermore, there is also no quantitative analysis on the relationships between the PLD and the free market prices with the regulated, auctioned prices for Small Hydropower plants. Since 2004, the A-5/A-3 nominal prices of small hydropower have been growing close to inflation. However, there is very little knowledge on what will drive their behavior on the future. With the current drought, investors are more inclined towards the free and the spot markets. Should the drought come to an end, and should the PLD return to the low-end levels, it is very unclear that the regulated prices would continue to maintain their value in real terms. Since there is complete data on auction prices and on the spot prices, even with no data on bilateral agreements there is already very significant space for study within this matter.

The fifth topic is regarding the impact of the governmental interventions in the electricity market. For instance, the cap created by ANEEL for the PLD destroyed significant value for investors who were leaving energy to be sold at the clearing/spot market. However, it did benefit plants that were not being able to deliver the energy they sold in advance, specifically those out of the MRE. Since the Brazilian Electricity Market structure was defined in 2003-2004, it has been fairly stable – in other words, the game did not change, but some key interventions did happen and affected players' performance. Public policies are usually based on good intentions, but often they provoke instability and the true message sent to investors is more of volatility than of protection. Citing a passage of the book "Investment Under Uncertainty", by Pindyck and Dixit, demonstrates well this behavior through commenting on interest rates: "if an objective of public policy is to stimulate investment, the stability of interest rates may be more important than the level of interest rates. Policies that lead to lower but more volatile interest rates could end up depressing aggregate investment spending." (Pindyck & Dixit, 1994)

Following on interest rates, the last topic proposed here is regarding the study of how investors determine the rate of annual return that they require. As explained in section 4.3.1.1., the model developed for this thesis utilized a rate of return in line to what was observed from investors. However, it did not explore the matter of how it was determined. Miscalculations or the utilization of sub-optimal rates of return may lead to significant loss of value through also sub-optimal investment decisions. As observed with the 8 portfolios utilized in the simulations, investing in Small Hydropower can produce revenues with very low standard deviations. Furthermore, the operations of this type of plants is nothing too complex, and the main value driver in this sense would be the capital expenditure and the construction process. Thus, it is an investment of very low risk, which could allow investors to have lower required rate of return than in other fields, even within a high-interest-rates country like Brazil.

Lastly, those topics were listed due to their direct observations on this thesis and their immediate relevance for the market. They are not exhaustive in terms of what still needs to be studied to provide better tools for players on the Brazilian Electricity Market. However, further development of knowledge in such fields may influence investors and regulators' actions, and thus may cause great positive impact on the market and on the country.

8. References

ABRACEEL. (2014). Retrieved March 13, 2015, from ABRACEEL Website: http://www.aneel.gov.br/aplicacoes/consulta_publica/documentos/Abraceel%20CP%20 009_2014.pdf

ANEEL. (1998). *Resolution Number 393.* ANEEL - Agência Nacional de Energia Elétrica. Retrieved April 3, 2015, from http://www.aneel.gov.br/cedoc/res1998393.pdf

ANEEL. (2004-2015). *Auction Conclusion Reports for Various Auctions.* ANEEL - Agência Nacional de Energia Elétrica. Retrieved March 12, 2015, from http://www.aneel.gov.br

ANEEL. (2010, August 10). *Resolução Normativa de Número 409.* Retrieved from Agência Nacional de Energia Elétrica: http://www.aneel.gov.br/cedoc/ren2010409.pdf

ANEEL. (2013, November 26). Retrieved April 4, 2015, from ANEEL Press Release: http://www.aneel.gov.br/aplicacoes/noticias/Output_Noticias.cfm?Identidade=7533&id _area=90

ANEEL. (2014, December 22). Retrieved March 13, 2015, from ANEEL News: http://www.cmuenergia.com.br/site/Noticia/ANEEL_reduz_preco_teto_do_PLD_para_2 015/296

ANEEL. (2014, December 10). Retrieved March 13, 2015, from ANEEL Press Release: http://www.aneel.gov.br/aplicacoes/noticias/Output_Noticias.cfm?Identidade=8297&id _area=

ANEEL. (2015). Retrieved March 17, 2015, from ANEEL - Agência Nacional de Energia Elétrica Website: http://www.aneel.gov.br/area.cfm?idArea=635&idPerfil=5

ANEEL. (2015). *BIG - Banco de Informações de Geração.* ANEEL - Agência Nacional de Energia Elétrica. Retrieved February 17, 2015, from http://www.aneel.gov.br/aplicacoes/capacidadebrasil/GeracaoTipoFase.asp?tipo=5&fas e=3

ANEEL. (2015). *Resumo Geral das Usinas – February/2015.* ANEEL - Agência Nacional de Energia Elétrica.

ASBRAV. (2014). *Associação Sul Brasileira de Refrigeração, Ar Condicionado, Aquecimento e Ventilação.* Retrieved April 29, 2015

BBCE. (2015). Retrieved March 22, 2015, from Balcão Brasileiro de Comercialização de Energia: http://www.bbce.com.br

BP Global. (2015, February 17). *BP Statistical Review of World Energy June 2014.* Retrieved from http://www.bp.com/statisticalreview

Brazilian Federal Government. (2011, September 15). Retrieved from Brazilian Government Website: http://www.brasil.gov.br/infraestrutura/2011/09/expansao-da-energia-nuclear-no-brasil-se-limita-a-angra-3-diz-tolmasquim

Brazilian Government. (1998, May 27). *LEI Nº 9.648, DE 27 DE MAIO DE 1998.* Retrieved from http://www.planalto.gov.br/ccivil_03/leis/l9648cons.htm

Canal Energia. (2015, March 18). Retrieved from Canal Energia: http://www.canalenergia.com.br/zpublisher/materias/Retrospectiva.asp?id=90762&a=2012

CCEE. (2001-2015). *PLD - Preço de Liquidação de Diferenças.* CCEE - Câmara de Comércio de Energia Elétrica. Retrieved March 20, 2015, from http://www.ccee.org.br/portal/faces/pages_publico/o-que-fazemos/como_ccee_atua/precos/precos_csv

CCEE. (2004-2015). *Reports of Results of Various Auctions.* CCEE - Câmara de Comércio de Energia Elétrica.

CCEE. (2013). *MRE - Mecanismo de Relocação de Energia.* Retrieved April 4, 2015, from MRE - Mecanismo de Relocação de Energia: http://www.ccee.org.br/ccee/documentos/CCEE_076159

CCEE. (2015). Retrieved March 17, 2015, from CCEE - Câmara de Comércio de Energia Elétrica: http://ccee.org.br/portal/faces/pages_publico/quem-somos?_adf.ctrl-state=17blfxn2nz_4&_afrLoop=17907087043988

CCEE. (2015). *Infomercado 2015.*

EIA-USA. (2015). Retrieved March 11, 2015, from Energia Information Administration Website: http://www.eia.gov/totalenergy/data/annual/#electricity

Enfoque. (2005). *Processo de Comercialização de Energia.* Retrieved from http://www.enfoque.com.br/infocias/infocias_doc/017329/05/ian01732912312005400. doc

EPE. (2007). *PEN 2030 - Plano Energético Nacional.* EPE - Empresa de Pesquisa Energética.

EPE. (2014). *Anuária Estatístico de Energia Elétrica.* EPE - Empresa de Pesquisa Energética.

EPE. (2014). *Balanço Energético Nacional.* EPE - Empresa de Pesquisa Energética.

EPE. (2015). Retrieved March 17, 2015, from Empresa de Pesquisa Energética Website: http://epe.gov.br/quemsomos/Paginas/default.aspx

Estado de São Paulo. (2014, March 15). Retrieved March 10, 2015, from Jornal o Estado de São Paulo: http://economia.estadao.com.br/noticias/geral,eletrobras-perde-r-19-bi-na-bolsa-e-ate-muda-de-sede-para-economizar-o-aluguel,179703e

Exame Magazine. (2013, December 13). *Exame Magazine.* Retrieved April 2, 2015, from http://exame.abril.com.br/economia/noticias/edp-e-furnas-vencem-disputa-por-sao-manoel-no-leilao-a-5

G1 News. (2015, March). Retrieved March 17, 2015, from G1 News: http://g1.globo.com/sao-paulo/noticia/2015/03/nivel-do-sistema-cantareira-sobe-de-137-para-14.html

G1 News; ANEFAC. (2015, March 11). Retrieved April 6th, 2015, from G1 News: http://g1.globo.com/economia/seu-dinheiro/noticia/2015/03/juro-medio-para-pessoa-fisica-sobe-para-66-em-fevereiro-diz-anefac.html

IBGE. (1994-2015). *IPCA - Índice Nacional de Preços ao Consumidor.* IBGE - Instituto Brasileiro de Geografia e Estatística.

IBGE. (2010). *Censo 2010.* Brasília: Instituto Brasileiro de Geografía e Estatística.

IBGE. (2015). *IBGE – I.51 Produto Interno Bruto.* IBGE - Instituto Brasileiro de Geografia e Estatística.

IMF. (2015). *IMF World Economic Outlook April/2015.* International Monetary Fund.

McKinsey and Company. (2014). *Global GHG Abatement Cost Curve.*

Melo, E., Neves, E. M., da Costa, A. M., & Correia, T. B. (n.d.). A Perspective of the Brazilian Electricity Sector Restructuring: From Privatization to the New Model Framework.

MME. (2004). *MME - Proinfa.* Retrieved April 29, 2015, from http://www.mme.gov.br/programas/proinfa/

MME. (2015). Retrieved March 11, 2015, from Ministério de Minas e Energia Website: http://mme.gov.br/web/guest/acesso-a-informacao/institucional/o-ministerio

National Weather Service. (2015, April 26). *Climate Prediction Center.* Retrieved from http://www.cpc.ncep.noaa.gov/

ONS. (2015). *Operador Nacional do Sistema Elétrico.* Retrieved March 10, 2015, from http://www.ons.org.br

Pérez-Arriaga, J. (2013). *Regulation of the Power Sector.* Madrid: Springer.

Pindyck, R. S., & Dixit, A. K. (1994). *Investment under Uncertainty.* Princeton University Press.

PortalPCH. (2014, October 9). Retrieved March 12, 2015, from PortalPCH Website: http://www.portalpch.com.br/noticias-e-opniao/4167-09-10-2014-custo-da-energia-a-vista-pode-cair-50-em-2015-valor-economico.html

Small Hydropower Investors, A. (2015, January). Interviews on Small Hydropower investment behavior. (T. R. Cortes, Interviewer)

The World Bank. (2015). *Data - The World Bank*. Retrieved April 29, 2015, from http://data.worldbank.org/indicator/SP.URB.TOTL.IN.ZS

VEJA Magazine. (2013, May 12). *VEJA Magazine*. Retrieved April 2, 2015, from http://veja.abril.com.br/noticia/economia/custo-da-usina-de-belo-monte-ja-supera-os-r-30-bilhoes/

XPGestão. (2014). *Quarterly Report - 2014Q3*. Retrieved March 16, 2015, from http://www.xpgestao.com.br/documentos/pdf/relatorios/2014/relatorios-fii/Relatorio%20XPOM11_201409_Trimestral.pdf

ABOUT THE AUTHOR

Thomaz Cortes has experience as a McKinsey & Company consultant, founding a Brazilian startup bringing solar water heating to low-income communities and working in Venture Capital. He also received his Master of Business Administration from the Massachusetts Institute of Technology Sloan School of Management. Thomaz graduated in Energy and Automation Engineering at the University of São Paulo.

www.ingramcontent.com/pod-product-compliance
Lightning Source LLC
Chambersburg PA
CBHW050723180526
45159CB00003B/1111